THIS IS SERVER COUNTRY

Essential Edition

THIS IS SERVER COUNTRY

AI, Power, and the Remaking of Rural America

ESSENTIAL EDITION

MICHAEL J. BOMMARITO II

2026

This Is Server Country:
AI, Power, and the Remaking of Rural America
Essential Edition

Copyright © 2026 Michael J. Bommarito II.

All rights reserved. No part of this publication may be reproduced, distributed, or transmitted in any form or by any means, including photocopying, recording, or other electronic or mechanical methods, without the prior written permission of the author, except for brief quotations in critical reviews and other noncommercial uses permitted by copyright law.

Abridged from *This Is Server Country*. For the complete work with full source documentation, see the unabridged edition.

Cover design and typesetting by the author. Printed in the United States of America.

Book Website: `servercountry.org`
Author Website: `michaelbommarito.com`

	Paperback	*Ebook*
Essential Edition	979-8-9943457-5-7	979-8-9943457-4-0
Full Edition	979-8-9943457-3-3	979-8-9943457-2-6

Publisher's Cataloging-in-Publication Data
Names: Bommarito, Michael J., II, author.
Title: This is server country – essential edition :
    AI, power, and the remaking of rural America / Michael J. Bommarito II.
Description: Essential edition. | Michigan : 2026. | Includes bibliographical references.
Identifiers: ISBN 979-8-9943457-5-7 (paperback) | ISBN 979-8-9943457-4-0 (ebook)
Subjects: LCSH: Data centers—United States. | Artificial intelligence—Economic aspects—United States. |
    Electric power—United States. | Infrastructure (Economics)—United States. |
    Rural development—United States.
Classification: LCC TK5105.85 .B66 2026 | DDC 004.6—dc23

To our children:
may you live a better life for these decisions we make.

Author's Note

As I finish this *Essential Edition*, January of 2026 is rapidly approaching its close. By the time you read this book, the facts will have changed. The issues, however, will no doubt remain.

Over the last seven months, I have documented more than 600 data center projects and countless policies across all fifty states. Together, they represent over a trillion dollars in announced investment and the most significant demand for power capacity in the history of our country. The database captures the eight largest technology companies driving demand, the sixteen major companies pouring concrete, the eleven primary sponsors providing money to finance the builds, and the many utilities and other stakeholders caught in the middle.

Just this month, a new 345 kV transmission route has been proposed that would condemn the houses of friends and run just a few hundred yards from my century-old bank barn. The feverish urgency with which I have written thus needs little explanation.

Some projects I have tracked are operational. Others exist only as construction sites and permits. Many will never pour a yard of concrete. Thankfully, I have accepted the inevitability of a second edition of these books. The very nature of this issue requires updates. I will post corrections, data updates, and additional resources at `servercountry.org`.

This Essential Edition condenses my core narrative and arguments into less than half the original length. The full edition explores the AI data center buildout through detailed qualitative and statistical discussion, extended case studies, and comprehensive source documentation. This edition distills that record to its essentials: the key scenes, the central arguments, and only the numbers that matter most.

For those who want the complete record with proper bibliographic references, the unabridged *This Is Server Country* edition is available in print and digital formats wherever you find this Essential Edition.

We have been here before. The railroads opened the West. The interstates created the suburbs. Each time, we built the future with concrete and steel. Each time, benefits flowed broadly while costs concentrated locally. The AI buildout follows the same pattern, at comparable scale, at faster speed.

But a database is not a story. To understand the machine, you have to stand in one place and watch it work.

In October 2025, Governor Whitmer announced that OpenAI and Oracle would build a $7 billion campus in Saline Township, Michigan. Saline is not the largest project in my database, nor the most controversial. But it is representative. The same forces acting on this Michigan township—power constraints, state competition, community resistance—are reshaping Virginia, Texas, and Arizona. This is a book about America, not just Michigan.

Michigan also happens to be the place I know best.

My wife and I grew up in mid-Michigan, returned when we started a family, and now live an hour from where this story begins. We earned five degrees between us at the University of Michigan, fifteen miles from the Saline site. We have a small farmstead. I have read the *Michigan Farm News* at home for years; that is where I first saw these issues stirring, long before they appeared in my FT subscription.

I also understand the other perspectives. I lead the ALEA Institute, a nonprofit focused on responsible AI—training open models, releasing open data, conducting policy research. I was part of the research team that tested GPT-4 against the bar exam. I even worked, once upon a time, in Manhattan for hedge funds.

More pointedly, I used AI to help write this book. Claude Code, Anthropic's agentic command-line tool, assisted with research, drafting, and revision. I rent an NVIDIA GH200 from Lambda Labs for my academic research—the same class of hardware filling the racks I describe. Every query or training step consumed compute cycles in data centers like the ones in these pages. The irony is not lost on me. I am complicit in what I document.

I am no main character in this story, but I hope you trust that I speak from experience as well as conviction.

I called this book *This Is Server Country* because the phrase echoes something you see on roadside signs across rural America: *This is God's Country, This is Trump Country*—declarations of identity and providence.

The AI buildout is making a similar claim, written in concrete and copper instead of paint and prayer. Whether that transformation is progress or loss depends on who you ask. I have tried to hold both futures in superposition together.

One question in particular runs through this book: why farmland and forest? America has thousands of shuttered factories, abandoned malls, and decommissioned power plants—sites already scarred by industry, already connected to roads and utilities. Why does the AI buildout sacrifice virgin soil and ancient oak instead? I came to this project assuming the answer was greed or indifference. The truth turned out to be more structural, more frustrating, and more worth understanding.

Some things I could document directly. Others, I had to help you imagine.

To do that, some characters in this book are composites—fictional people representing documented roles: *Frank, Ellen, Harold, David, Steve, Roberto, Sarah, Elena, Wei-Lin,* and others. Their dialogue and inner thoughts are invented, though grounded in trade publications, public testimony, and news accounts. You will recognize these composites by their italicized single names. Where real people appear by full name—governors, executives, regulators—their words come from the public record.

This is a book about AI, power, and politics: the silicon that demands electricity, the land that hosts it, and the deals—from township zoning to trade policy—that determine where it gets built. It does not address what AI might eventually become—whether these systems will remain tools or become something else entirely. That question deserves its own book.

It does not argue that data centers are good or bad. It argues that they are consequential. The buildings going up now will stand for decades. The communities hosting them will be transformed whether they chose transformation or not. The farmland beneath them will not return.

The choices being made now deserve more scrutiny than they are getting. This book is my attempt to pay attention before the concrete sets and the soil is gone.

—MJB, January 2026

Contents

Author's Note. v
Prologue: The Token 1
1   The Inference 13
2   The Silicon 23
3   The Data Center 31
4   The Power Constraint 43
5   The Grid. 51
6   The Generation. 61
7   The Land 71
8   The Money. 85
9   The Incentives 99
10   The Geopolitics. 111
11   What Could Change 123
12   Our Future 133
Epilogue: The Token, Revisited 149
Selected Sources. 157

PROLOGUE
The Token

You type a question. A second later, words appear. The cursor blinks. More words arrive, assembling themselves into sentences, paragraphs, thoughts. You barely notice the miracle.

It is a Tuesday evening in January 2026. You are sitting in a coffee shop on Liberty Street in Ann Arbor, Michigan. The winter dark has already settled in. Through the window, students trudge past in heavy coats, their breath visible in the cold. Your laptop glows on the table beside a cooling cup of coffee.

The response appears at a pace that mimics typing. But no one is typing. No human sits on the other end of this conversation. The words emerge from a process so complex, so distributed, that even the engineers who built it cannot fully explain how it works.

Between your fingertips and that answer, a vast infrastructure hums. Fiber optic cables carry your question at the speed of light. Servers parse your words into tokens—numerical fragments that neural networks can process. And somewhere, in a building most people will never see, electricity flows through silicon at rates that would have astonished the engineers who built America's first power plants.

Fifteen miles southwest of where you sit, the answer takes physical form. A facility nicknamed "The Barn" is rising across 250 acres of former farmland in Saline Township. Three buildings, each the size of five football fields. When fully operational, this complex will consume 1.4

Prologue

gigawatts of electricity—roughly the output of a large nuclear reactor. Enough to power a city of 800,000 people.

Your question consumes a sliver of electricity. A few watt-hours, perhaps a penny's worth of power. Trivial. But you are one of hundreds of millions asking questions today. And that chat window is only what you see. AI inference runs beneath the surface of modern life. It shapes the search results that now include AI summaries. It autocompletes your emails. It decides what appears in your social media feed and recommends what you watch next. A typical American triggers thirty to a hundred AI inferences per day, often without knowing it. Power users, developers with coding assistants and professionals with AI-augmented workflows, may trigger hundreds. The variation is enormous, the accounting imprecise. No one tracks the total. And that is just individuals; enterprises and governments run millions more through APIs that never touch a chat interface.

Multiply these invisible queries across a population, and the slivers become a torrent. ChatGPT alone processes 2.5 billion prompts daily—and ChatGPT is one product among dozens. The infrastructure exists not for any single question but for the staggering volume of them, every hour of every day, from every corner of the world. Seven billion dollars in construction. 1.4 gigawatts of continuous power. All to serve the collective appetite for answers.

The token that appears on your screen, each fragment of a word, represents electricity transmuted into something that resembles thought. Every token is cognitive labor, produced at industrial scale, at a cost no human worker could match. The infrastructure rising across America is not just serving curious students. It is converting capital into cognition—and reshaping the economy in ways we are only beginning to understand.

Saline Township sits in Washtenaw County, about fifteen miles southwest of Ann Arbor. Farm country. Flat fields and two-lane roads, scattered houses on large lots, 2,300 people spread across nearly thirty-five square miles. Downtown has a veterans' memorial in Oakwood Cemetery and City Limits Diner, where farmers have had coffee in the same booths for forty years. Henry Ford rebuilt the dam in 1936 to power a soybean plant that bought crops from hundreds of local farmers. The plant closed decades ago. The dam remains.

The township government is modest: a five-member board, a small budget, a volunteer fire department. Until 2025, the biggest controversies involved zoning variances for home additions. Board meetings attracted a handful of regulars. Decisions were made without fanfare.

Frank had been township supervisor for eight years. A retired auto industry engineer, he ran for the board after getting annoyed about a drainage issue near his property. The job paid a modest stipend and consumed evenings that might otherwise go to his grandchildren—*Lily*, eight, and *Marcus*, six, who lived twenty minutes away in Ann Arbor and asked every Sunday why Grandpa couldn't come to soccer. His wife asked the same question, in different words. But he took the job seriously. Someone had to.

In January 2025, nothing in his experience prepared him for what came next.

In early 2025, a developer called Related Digital began quietly assembling parcels of farmland along Michigan Avenue. The road has another name: US-12, the old Chicago Road. Pioneers cut it through the wilderness in 1827, following Indian trails west. For a century it was the main route between Detroit and Chicago, and Saline grew up along it, serving travelers and farmers heading toward a city that barely existed.

Prologue

Then I-94 opened in 1961. The interstate rendered the old road a relic. Some farms along it have been in the same families for over a hundred years. Metal plaques—some donated by the local utility—mark the survivors: Centennial Farms.

That developer assembling parcels along Michigan Avenue was not a local operation. Related Digital is the data center arm of Related Companies, founded by Stephen Ross—the billionaire who owns the Miami Dolphins and whose name adorns the business school at the University of Michigan, fifteen miles up the road in Ann Arbor, and whose ties to the state run deeper still. The people who built Hudson Yards in Manhattan had come to Saline to capitalize on the AI boom.

The company filed for rezoning without revealing who would occupy the site. Residents noticed the activity, and questions arose at board meetings. Representatives spoke of "advanced technology" and "high-wage jobs" but deflected specifics. Who was the tenant? They could not say. How much power would the facility consume? "Significant." What about water usage, traffic impacts, noise levels? Vague assurances.

The Planning Commission had already rejected the proposal in August. Now, on September 10, 2025, the full Board would vote. Sixty residents packed the Saline Township hall—more than *Frank* had ever seen at a board meeting, though fewer than the developer had probably expected. Outside, the evening had turned crisp, the kind of September night that reminded you summer was over. Inside, the room grew warm with bodies and voices.

The public had questions. Traffic. Water. Pollution. Sound. What happens when the power grid fails? What happens when the company sells? *Ellen*, whose family had farmed adjacent land since 1947, spoke about groundwater concerns. She was sixty-six years old, had taught biology at Saline High School for thirty years before retiring, and could identify every bird that crossed her property by its song. Her hands shook as she read from handwritten notes. "My grandfather dug that well with his own hands," she said. "Fifty-two feet deep, through clay and sand, with a shovel and a mule-drawn pulley. You're asking us to trust that a company we've never heard of will monitor it, fix it if it runs dry, and honor that promise twenty years from now. What happens when they sell to someone else?"

The developer's attorney responded with slides and statistics. Projected employment: 450 permanent positions. Tax revenue increases: 400 percent. Environmental monitoring: comprehensive. *Frank* found himself wondering how many township boards this lawyer had addressed, with how many variations of the same presentation.

After the testimony concluded, *Frank* called for the vote. He raised his own hand before fully processing that he was raising it. Four to one to deny the rezoning request. Board members cited concerns about fire department preparedness, conflict with the township's master plan, and inadequate answers to basic questions. One member called the proposal "a self-imposed environmental catastrophe."

The room erupted in applause. *Ellen* hugged the woman next to her. Someone shouted, "That's democracy!"

In the back, the developer's team was already packing up their materials, their faces carefully neutral. They didn't look surprised, *Frank* noticed. They didn't look concerned. They looked like people who had been through this before and knew what came next.

Afterward, in the parking lot, *Ellen* found him by his truck. Her hands were still shaking, the way they had been at the podium. "Thank you," she said. "I wasn't sure which way you'd vote."

"Neither was I," *Frank* admitted. "Not until my hand was up." He looked back at the community center, where the lights were going dark. "*Ellen*, I want to be honest with you. This isn't over. You saw their faces."

"I know." She pulled her cardigan tighter against the September night. "But at least we tried. At least we went on the record." She paused. "*Harold* wasn't here tonight."

Prologue

Frank had noticed. *Harold*, who had farmed the land next to *Ellen*'s since they were both young, who had lost his wife two years ago and whose children had long since moved away. *Harold*, whose five hundred acres lay directly in the developer's path. "I tried calling him yesterday," *Frank* said. "He didn't pick up."

"He's been avoiding everyone. I brought him casserole last week—he barely opened the door." *Ellen* looked toward the dark fields beyond the parking lot lights. "I think they've already made him an offer. A big one."

Neither of them said what they were both thinking: that the applause inside might not matter, that democracy had limits, that *Harold*'s decision was his alone to make.

Two days later, Related Digital filed a lawsuit.

The lawsuit alleged that the township's denial was arbitrary and capricious. It sought to overturn the board's decision through the courts.

The township faced a dilemma. Defense would cost hundreds of thousands of dollars in legal fees—and the township's entire annual budget was modest by suburban standards. A prolonged fight could drain the treasury. Even if the township prevailed at trial, the developer could appeal, stretching litigation across years and mounting costs.

Related Digital had essentially unlimited resources—backed not just by Stephen Ross's billions but by the entire Stargate initiative, the five-hundred-billion-dollar AI infrastructure project announced by President Trump earlier that year.

The township's lawyers advised the board that winning was uncertain. Michigan courts often defer to property owners seeking to develop their land. And even if the township won, the victory could be Pyrrhic. By the time the litigation concluded, years later, the township would be financially exhausted.

Frank spent the first week after the lawsuit unable to sleep past 4 a.m. He would lie awake in the dark bedroom, listening to his wife's steady breathing, watching the digital clock on the nightstand mark the hours in red numerals. 4:17. 4:23. 4:41. The house made its familiar sounds—the furnace cycling on, a branch scraping the gutter, the refrigerator's distant hum—and none of it was comforting anymore. The ceiling above him was the same ceiling he had stared at for thirty-two years in this house, but it looked different now. Heavier.

The township's annual budget was less than Related Digital would spend on lawyers in a month. He thought about *Ellen*, who had thanked him after the September meeting with tears in her eyes, gripping his hands like he had saved something precious. He thought about *Lily* and *Marcus*, who would inherit whatever decision he made. *Marcus* had asked him at dinner last Sunday what a data center was, and *Frank* had tried to explain, and the boy had said, "So it's like a really big computer?" Yes, *Frank* had said. It's like a really big computer that needs all the electricity in the world. The boy had nodded and gone back to his macaroni, satisfied. *Frank* had not been able to finish his own dinner.

On September 24, the board held a special meeting. The soybeans along the highway had turned yellow, ready to cut. The corn stood tall and dry, waiting. In a few weeks the combines would come through and the fields would be bare. Then winter. Then spring. Then someone else's problem.

David Landry addressed the board and the thirty residents who had come to hear the township's options. Landry was a municipal attorney from Farmington Hills, a former mayor of Novi who had spent four decades handling cases like this one. The township's insurance company had sent him. He laid out the facts without sugarcoating them.

"You can fight this," Landry said. "But you need to understand what fighting means. Two to three years of litigation. Depleting the township's reserves. The possibility of losing anyway and paying their legal fees on

Prologue

top of yours. And the project gets built eventually regardless, just with a worse relationship and no concessions."

The options were settle, negotiate a consent judgment, or go to court. Michigan courts often deferred to property owners. The township would have to prove the project would cause severe harm to infrastructure. The studies the developer had submitted made that difficult.

"It's not what your residents wanted," Landry said. "But negotiating might be the best outcome available."

He made clear he was not telling the board how to vote. That was their decision. He was there to explain the law, not to make the choice for them.

Within three weeks of the lawsuit being filed, the board met again. The room was smaller this time—word had spread that the deal was done. *Frank* read a statement explaining the board's reasoning. His voice cracked once, on the word "impossible."

Then the board voted four to one to settle the lawsuit and approve the project. One member who had voted to deny now wanted to keep fighting—but the math had changed, and she was outvoted. The consent agreement included concessions: well monitoring, fire department funding, noise limits, a community investment fund. It was not what anyone had wanted. It was what they could get.

Local democracy spoke on September 10. Something larger answered on September 12. By October 1, the matter was settled.

Only after the settlement did the full picture emerge. In late October, Governor Gretchen Whitmer stood in Saline Township and announced "the largest single investment in Michigan history." The anchor tenant was OpenAI, the maker of ChatGPT, in partnership with Oracle.

The Michigan Public Service Commission approved the power contracts in December 2025. More than 5,500 public comments were submitted, community leaders voiced overwhelming opposition, and public officials from both parties urged the Commission to slow down. The Commission approved them anyway, without holding a contested hearing. Attorney General Dana Nessel criticized the process, noting that key contract details remained redacted even from her office.

Saline Township is not unique. Across America, similar confrontations play out. In Augusta Township, just east of Saline, a Google data center proposal faces a referendum. A massive project called the Digital Gateway was defeated in Prince William County, Virginia, after years of litigation—twenty-four billion dollars in investment, rejected. Google abandoned a billion-dollar proposal in Indianapolis after opposition intensified. Microsoft walked away from a project in Caledonia, Wisconsin.

But these victories are exceptions. For every Digital Gateway that falls, a dozen Saline Townships acquiesce. The pattern that emerged here—rejection, lawsuit, settlement, approval—offers a template. Developers with deep pockets can outlast local governments with shallow ones.

Between 2024 and 2030, technology companies, private equity firms, and sovereign wealth funds plan to invest more than one trillion dollars in data center construction across the United States. Over 600 projects. More than 131 gigawatts of planned power capacity—roughly seventeen percent of current American electricity consumption, concentrated in a few hundred facilities. Microsoft alone has committed eighty billion dollars per year. Amazon, Google, and Meta each spend tens of billions more.

We are building the largest private infrastructure project in American history. Most Americans know nothing about it.

Prologue

Here is the central insight of this story: the constraint on artificial intelligence is not software, not algorithms, not data, not even the specialized chips that perform the calculations. The constraint is electricity.

The companies building AI need electricity at a scale American utilities have never supplied to single customers. A gigawatt was once the output of the largest power plants; now it is the requirement for a single data center campus. That power must come from somewhere, and new generation takes years to build. Transmission lines must carry it, and upgrading those lines takes even longer—often a decade or more.

These requirements determine where data centers can be built more than any other factor—not land prices, not labor costs, not proximity to customers. Grid access drives everything. It explains why data centers cluster in Northern Virginia, where decades of infrastructure created the necessary capacity. It explains why new projects target rural areas, where transmission lines built to serve power plants can be repurposed for consumption. And it explains why a township of 2,300 people suddenly found itself at the center of a seven-billion-dollar deal.

This book traces the infrastructure that makes your ChatGPT query possible. We work backward from the token on your screen to the geopolitics that shape where data centers get built. The journey runs through inference mathematics, silicon chips, data center design, grid topology, power generation, land economics, capital flows, and political deal-making. Throughout, we return to Saline Township—not because it is the largest project or the most controversial, but because it offers a window into all the forces at play.

Stand in Saline Township today and you see construction equipment, security fencing, and a preserved red barn that gives the project its nickname. The barn is a sentimental touch, kept because it photographs well. It will remain while the fields around it become something else entirely.

The decisions being made now will structure possibilities for decades. The precedents being set—about power consumption, land use, environmental impact, and local control—will apply to hundreds of future projects. These trade-offs will shape what kind of communities Americans live in and what resources remain available to them.

The winners are celebrating. The losers are just beginning to understand what they lost.

Frank still lies awake some nights, the same thoughts cycling through his mind: The township's budget. The developer's lawyers. *Ellen*'s well. *Lily*'s face when he missed her birthday party for a board meeting. *Marcus* asking if the big computer was finished yet.

He drives past the construction site sometimes, on his way to nowhere in particular. The red barn stands in the middle of it all, preserved because someone decided it photographed well. He remembers when it was just a barn.

CHAPTER ONE

The Inference

At 2:47 a.m. on a Tuesday in January 2026, a college student in Ann Arbor sits at her desk, laptop glowing in a dark dorm room. She has a biochemistry exam in six hours and three chapters left to review. She types into ChatGPT: "Explain how the Krebs cycle generates ATP, like I'm cramming for an exam." She hits Enter.

For a few dozen milliseconds, nothing happens. Her request travels through campus fiber, crosses into a carrier's backbone, and lands at a load balancer in a cloud data center. The system checks who she is, checks what she is allowed to do, and decides where to send her question. Then the response begins: words appear on her screen, one by one, in a rhythm that mimics typing.

Fifteen miles southwest, a facility is rising from Michigan farmland—chosen not for its university or workforce, but for the high-voltage transmission lines that cross it. Within a year or two, requests like hers may route through Saline Township instead of Virginia or Texas. The interface will look the same. The physics will not.

Her question becomes tokens—numerical representations of text, split into words and word fragments. "Biochemistry" might become four tokens: "bio," "chem," "is," "try." The model never sees letters. It sees numbers. Those numbers enter a neural network, passing through layer after layer of computation. No database lookup. No internet search. Instead, something both simpler and more unsettling: the model predicts the next token.

Given everything in the prompt, what should come next? A probability distribution emerges. The model samples from it, chooses a token, appends it to the prompt, and repeats. Token by token, the response builds. That is why the answer streams—each new word depends on all the ones before it. Even with caching, work accumulates. The three-hundredth token is not free; it sits atop the previous two hundred and ninety-nine.

This process is called inference—the atomic unit of modern AI work, the operation that makes ChatGPT, Claude, Gemini, and a thousand quieter systems useful. The student knows none of that. She only knows the answer is clear and practical, exactly what she asked for. She highlights a paragraph, copies it into her notes, asks a follow-up. Another dozen tokens in, a few hundred out, another half-second of computation somewhere in America.

The scale of the infrastructure is hard to grasp because the experience feels intimate. She feels like she is talking to something, not using something. The response matches her tone: quick and practical, not academic. That is the public relations miracle of this interface—the cost vanishes. We type into a text box. We do not see the GPUs, the cooling towers, the transmission lines, the land being rezoned in townships we have never visited. But the work still happens.

OpenAI's systems process billions of requests per day. Each request demands compute, consumes electricity, generates heat that must be removed. Add those slivers together across a day and the numbers stop looking like software. They start looking like industry. This is why the data center buildout is happening. A trained model is not the finish line—it is the starting point. That model must be served to hundreds of millions of people on demand, at speeds that feel instant. And then the next model, and the next. We have entered the inference era.

1.1 The Inference Era

To understand why inference dominates day-to-day cost, we have to separate it from training. Training creates a model—the long, expensive phase where parameters are adjusted by exposing a neural network to vast amounts of text. Parameters are the numbers inside the model that determine how it responds. A model with a hundred billion parameters has a hundred billion adjustable numbers. Training nudges those numbers toward better prediction. Arduous, but finite. Once complete, the weights get frozen and deployed. Those same weights serve every user, every request, again and again.

Inference, by contrast, never stops. Every time someone asks a question, the model springs into action. Weights load into GPU memory. Input passes through every layer. The response generates token by token. Then the weights wait for the next request. Think of it like writing a book versus reading it aloud. Writing is hard, but you do it once; the book exists. Reading that book aloud to every person who wants to hear it, individually, on demand, twenty-four hours a day—that is where work accumulates. Training is patient. Inference is impatient. Users do not accept "come back tomorrow."

So what is the model doing, exactly, when it "thinks"? At its core, a language model is a probability machine: given some text, it predicts what comes next. Type "The capital of France is" and the model does not consult a map. It performs a mathematical transformation on the input, drawing on patterns learned during training. Out comes a distribution over possible next tokens, and one token dominates: "Paris." Append "Paris" to the prompt and repeat. What comes next? Another distribution, another token, another repeat. This next-token loop is called autoregressive generation—the model feeds its own output back as input, generating one token at a time, like typing a sentence one letter at a time without being able to skip ahead. That loop is the heartbeat of modern language AI.

The transformer architecture, introduced in 2017, makes this loop powerful at scale. Its signature mechanism is attention: a way for every token to interact with every other token when the model decides what to do next. Consider a simple example: "The animal didn't cross the street because it was too tired." What is "it"? We know without thinking—the animal. Tired applies to animals, not streets. Attention is how the model makes that same connection. Processing the token "it," the model assigns higher importance to "animal" and lower importance to "street," doing this across many attention "heads," each learning different relationships. These outputs combine, pass through more layers, and produce the next distribution. A beautiful idea. Also expensive—dozens of layers per token, each layer reading the model's parameters and performing the transformation, repeated for every token in the response.

Here is what surprises people who have only heard "AI is math." The bottleneck in inference is not primarily computation—modern GPUs are astonishing at matrix multiplication. The bottleneck is memory. Large models carry enormous sets of weights, numbers that must be available quickly for every token. Those weights have to live close to the GPU cores, in high-bandwidth memory. Bigger models do not fit on one GPU. They split across many, adding coordination and overhead.

The cache adds another dimension. To avoid recomputing the same intermediate values over and over, the model stores a key-value cache as it generates tokens. That cache grows with the length of the conversation. Longer context windows help humans but cost more to serve—the cache can grow until it rivals the model itself. In practice, inference often becomes "memory-bound": the GPU waits for data to arrive, rather than waiting for math to finish. That distinction shapes everything downstream. Chip design. Data center design. The economics of producing a token.

1.2 The Hidden Stack

Zoom back out to the entire request. The student's question does not hit the neural network immediately. Before inference begins, the request passes through encryption, routing, authentication, rate limiting, safety filters, model selection, and scheduling. After inference, more systems engage—safety checks on the output, logging for billing and capacity planning, compression, streaming back to the browser. Not one program but a stack of services refined over years, designed to do something that sounds impossible: serve billions of requests with high reliability, without melting the GPUs or falling over under load.

Traditional software manages digital resources. Inference software adds the management of physical ones: power, cooling, networking, real estate. The student's question is computation. It is also heat, electricity, the physical world inside a building.

Scale alone does not explain the data center buildout. Latency does. If inference could happen anywhere, companies would consolidate compute where electricity is cheapest. They cannot. Users expect responses now, not later. For conversational tools, the key metric is time to first token: how long we wait before the answer starts streaming. Under a couple hundred milliseconds feels instantaneous. Over a second feels slow. Past a few seconds, many users assume something is broken. Code completion has even tighter tolerances. If suggestions lag, developers stop trusting the tool; the product stops feeling like an assistant and starts feeling like a distraction.

Physics has a vote here. A data center in Virginia cannot give a low-latency experience to a user in Tokyo. The speed of light imposes a floor on round-trip time. Serving global demand requires geographically distributed infrastructure. Compute has to be near us—not everywhere, but many places, including places like Saline Township.

Latency and cost pull against each other. Low latency means keeping GPU capacity in reserve so requests can start immediately. Expensive. High efficiency means batching requests so GPUs stay full. Cheaper, but slower. Production systems juggle this trade. Paid tiers get dedicated capacity. Free tiers get batching. Engineers tune scheduling and caching because a few percentage points of efficiency, at this scale, mean real money. But optimization has limits. Demand growth tends to outrun efficiency gains. The system gets better—and still expands.

Every token consumes electricity. Every joule becomes heat. Every watt of heat must be removed by cooling systems that themselves draw power. A single modern AI GPU draws roughly seven hundred watts when working hard—enough heat to warm a small room. Now picture thousands of them in one building, running all day and all night. Unlike many industrial processes, inference cannot shift to off-peak hours. Someone is always awake, always working, always asking. For utilities, that "always-on" demand is different. No breathing room. No quiet season. No easy maintenance window.

1.3 Whose Token?

Token generation looks different depending on where you stand. A machine learning engineer sees benchmarks: tokens per second per dollar, throughput under latency constraints. A one percent improvement can save millions at scale. A technology executive sees revenue and margin; capacity planning becomes a question of how many tokens can be served, at what cost, with what return. The student in Ann Arbor sees nothing. She cares about the answer, not the infrastructure.

But she is also a citizen, a voter, a ratepayer. The infrastructure serving her is being built somewhere, powered by a grid shared with someone else, enabled by policies funded by taxpayers. For residents of Saline Township, inference is not abstract. It is a transformation: farmland assembled into parcels, fences erected, trucks arriving, a new kind of neighbor moving in. Some see opportunity—construction work, permanent jobs. Others see disruption, risk, loss. Engineers optimize throughput. Executives optimize return. Users optimize convenience. Communities absorb externalities. That misalignment is a political problem, not a technical one.

Inference demand keeps rising for a simple reason: more and more things start to feel like a text box. Models improve, so people ask more of them. Existing products integrate AI, so millions of users go from zero requests a day to dozens, without ever choosing to adopt a "new" tool. Entire workflows rebuild around inference: claims processing, document review, support tickets, coding, writing. And as we move from chatbots to agents—systems that plan, act, check their work, and iterate—one user-visible response can hide many internal calls. The same interface multiplies the token load behind the scenes.

For the student asking about the Krebs cycle, these costs remain invisible. Her question arrives. An answer appears. The carbon emitted, the water consumed, the land transformed—those are borne by someone else, somewhere else. The tokens are not free.

1.4 Where It Lands

Return to Saline Township, where technical abstractions take physical form. Inside a modern AI facility, everything is organized in hierarchies: processors in sleds, sleds in racks, racks in rows, rows in halls, halls in buildings, buildings on a campus. That hierarchy matters because failures cascade. A dead component becomes degraded capacity. A dead rack becomes rerouted requests. A dead hall becomes a bad night for thousands of users. Redundancy is engineered at every level: multiple power feeds, backup generation, diverse cooling systems, multiple network paths. The goal is what the industry calls "five nines" availability—99.999 percent uptime, roughly five minutes of downtime per year—so reliable that users never think about it.

When the student types her question at 2:47 a.m., she assumes an answer will appear. That assumption rests on decades of reliability engineering, now applied to facilities that draw power like small cities. The human geography matters too. A facility like Saline will employ a few hundred permanent workers: electrical engineers, mechanical engineers, network engineers, security, operations. The machine learning engineers who design the models tend to work in expensive coastal cities; the operational jobs are where the buildings are. Construction employs far more people, but temporarily. Thousands of workers pour concrete and install electrical systems, then move on to the next job in the next state.

The student in Ann Arbor benefits. Engineers in San Francisco benefit. Investors in New York benefit. For now. Saline Township hosts the infrastructure and debates whether the jobs offset the transformation of its land. The question we keep circling: Is that exchange fair? Who gets to decide? And for how long do the winners stay winners?

Inference consumes enormous resources, but those resources take specific forms. This work does not run on ordinary processors. It runs on specialized chips: GPUs designed for massive parallel math and high memory bandwidth. These chips are not interchangeable. They are scarce. Production concentrates in a few places. Power demands shape the buildings that house them. To understand the inference era, we have to follow the hardware.

In the next chapter, we go down another layer, into the silicon.

CHAPTER TWO

The Silicon

TSMC is Taiwan Semiconductor Manufacturing Company, the world's leading chipmaker, producing roughly ninety percent of the most advanced AI chips. The clean room at Fab 18 operates in perpetual twilight, yellow-orange lights filtering out wavelengths that could ruin photosensitive wafers. At 3:17 a.m. Taiwan time, a process engineer named *Wei-Lin* watches a silicon wafer slide into a machine the size of a small house. Its sole job: print impossibly small patterns onto silicon. The machine costs hundreds of millions of dollars and takes years to deliver. Lasers and mirrors polished to atomic smoothness. Molten tin turned into a strobe so bright and precise that it sketches circuitry at scales our brains were not built to imagine.

Wei-Lin has learned the rhythm of the tool the way a pilot learns an aircraft. "One particle," he says, watching a sensor trace. "One contamination event. Millions of dollars gone." He has seen it happen. A speck of dust smaller than a cell lands in the wrong place and destroys a batch of chips that would have powered thousands of human conversations.

These chips are destined for America. Not for phones or laptops, but for data centers built around AI. Some will end up in places like Saline Township, Michigan, installed into racks that draw power like a row of houses. Here the token begins—not as a word fragment on a screen, but as etched silicon traveling through a global supply chain.

CHAPTER 2. THE SILICON

2.1 Why GPUs

To understand why a single chip can cost as much as a car, we have to understand what it does. Under all the branding, the work is basic arithmetic repeated at absurd scale: multiplication, addition, over and over. If we zoom in on a transformer model, most of the compute is matrix multiplication—large grids of numbers multiplied together, then summed. Do that enough times, fast enough, and we get something that resembles language.

Traditional CPUs power our laptops. They are designed for flexibility, handling many kinds of tasks well, one after another. A CPU is a smart generalist. A GPU—Graphics Processing Unit—is a specialist. The GPU emerged from video games, where millions of pixels must be computed in parallel, dozens of times a second. Each pixel is simple; what matters is doing many simple operations at once. That is exactly what modern AI needs, so the GPU, built to render dragons and race cars, becomes the engine of the inference era.

In 2012, a trio of researchers trained an image-recognition system called AlexNet on consumer gaming GPUs. The result was not a small improvement but a step change that kicked off the deep learning boom. NVIDIA, which had been a gaming company, suddenly found itself in the middle of a new arms race. It also had a quiet advantage: years earlier, NVIDIA had released CUDA, a programming framework that let developers use GPUs for general computing. When the deep learning wave hit, NVIDIA already had the tools, the libraries, and the developer habits. Hardware mattered. Software mattered just as much.

2.2 NVIDIA's Lock

By the mid-2020s, NVIDIA controls the overwhelming majority of the market for AI accelerators. No other company has held that kind of dominance in a sector this strategically important. How does that happen?

2.2. NVIDIA'S LOCK

Three interlocking advantages: purpose-built chips, sticky software, and control over supply.

Start with the silicon. NVIDIA's flagship AI chips are not generic processors that happen to be good at AI. They include specialized circuits called tensor cores, designed for the multiplication patterns that neural networks use. Each new generation arrives with a promise that everyone in the industry hears the same way: you can do more, faster, if you buy now. Every generation comes with a physical price. More performance means more power. More power means more heat. More heat means the end of air cooling. That thread ties a wafer in Taiwan to a liquid-cooled rack in Michigan—the silicon drives the building.

Memory is the invisible constraint. For inference, we often hit a limit not in raw math, but in how fast data can move. High-bandwidth memory exists for this reason: specialized memory chips stacked in layers and placed as close to the processor as possible. Advanced packaging becomes a bottleneck too—the craft of connecting multiple smaller chips and memory into one tightly coupled system. NVIDIA does not just sell chips now. It sells systems: tightly integrated platforms of GPUs, CPUs, memory, and networking designed to function as one unit. Inside a rack, fast interconnects link GPUs so they can behave like one larger machine. That architecture lets massive models run across many processors without falling apart into slow coordination overhead.

Hardware alone does not explain NVIDIA's dominance. Software plays an equally important role. CUDA is not just a language. It is a habit, a pile of optimized libraries, years of tooling and debugging experience baked into the daily work of millions of engineers. Even when competitors offer chips with impressive specifications, most customers hesitate. Switching means rewriting code, retraining staff, rebuilding pipelines, and taking performance risks in production systems that need to work all day, every day. "Better hardware" is rarely enough. A rival chip twenty percent cheaper but requiring months of migration work? The real cost is higher. The inertia is human.

NVIDIA also controls supply. The company does not manufacture its own chips—it designs them and outsources production to TSMC, the world's most advanced chip foundry. But NVIDIA decides who gets chips and who waits. When demand exceeds supply, as it did for years, allocation becomes power. Strategic partners get prioritized. Winners and losers are made with shipping schedules. In a shortage, your place in line matters as much as your budget.

2.3 The Challengers

So what about challengers? There are many. AMD offers powerful accelerators with lots of memory at lower prices, and some big players run major workloads on AMD chips. Cloud companies build custom chips for themselves—Google's TPUs, Amazon's Trainium and Inferentia—which can be excellent within their own clouds, but they come with their own software friction. China, constrained by export controls, pushes domestic alternatives. Some Chinese chips are good enough for many tasks today; the harder question is whether China can match the cutting edge without the same manufacturing equipment and supply chain. And startups like Groq (no relation to xAI's chatbot Grok) and Cerebras have built inference-focused hardware with architectures that look nothing like NVIDIA's, optimized for producing tokens quickly and predictably.

Some of these challengers are real. But the default choice holds—not because alternatives do not exist, but because NVIDIA has built a combined product: silicon plus software plus supply.

Hardware is only half the story. The other half is software: how efficiently models use the chips they run on. In early 2025, a different kind of threat arrived. Not from a chip company, but from an algorithm. A Chinese AI lab called DeepSeek released a reasoning model that matched or exceeded OpenAI's flagship model on certain key benchmarks—at a fraction of the training cost American labs had been spending. Constraint had bred innovation.

The lesson is not that hardware stops mattering. Efficiency is a moving target. If models get smarter per watt, does the infrastructure buildout shrink? History suggests the opposite—a pattern economists call Jevons paradox. When something becomes cheaper, people use more of it. Better engines led to more driving. Efficient lights led to more lighting. Cheaper computation led to more computation. In AI, cheaper tokens make new applications possible and drive higher usage. Efficiency gains become demand multipliers, not demand reducers. The buildout does not shrink—but efficiency gains inject uncertainty into its shape. Better algorithms can change where compute happens. Some inference shifts toward devices at the edge, even as total demand keeps climbing.

2.4 The Bottleneck

Even if demand is certain, supply is not. Every advanced AI chip depends on TSMC—not a casual dependence but a near-single point of failure for the most advanced silicon in the world. Taiwan is close to mainland China. Any disruption, whether natural disaster, military conflict, or political crisis, could halt the production of chips that underpin most of the AI economy. The United States tries to hedge with domestic manufacturing investments. Companies are building new fabs in Arizona, Texas, and Ohio. But a fab is not a warehouse. Construction takes years. High yields take longer. For now, the supply chain runs through Taiwan.

Geopolitics tightens the constraint further. Export controls limit who can buy the most advanced chips. Policy choices shape which countries get access to the best hardware—and which countries are forced into workarounds, smuggling, or domestic reinvention.

Talent is the final constraint. Only so many people on Earth can design a leading-edge GPU, optimize the software stack, build advanced packaging, and run a high-yield fab line. A handful of companies and countries hold that expertise. A small number of human beings decide what kind of AI hardware the world gets. That fragility cannot be fixed quickly, no matter how much money you throw at it.

2.5 Heat and Money

Pull the camera back to the data center floor. Chips get more powerful. They get hotter. Heat changes everything. For years, data centers were mainly a real estate story: find cheap land, build a box, push air through racks. That era is over. Rack densities have climbed from single-digit kilowatts to well over a hundred kilowatts for modern AI systems. At that density, air cooling fails. Data centers become thermal engineering projects: liquid cooling, coolant distribution, cold plates, sometimes immersion systems where servers sit in dielectric fluid. The cooling supply chain becomes its own bottleneck. In the old world, power and floor space meant you could install servers. In the new world, a facility with power and space but without liquid cooling hardware sits empty.

High density creates another challenge: networking. Training large models means thousands of GPUs exchanging data continuously. Slow network? Expensive compute sits waiting for packets. At cluster scale, the networking bill can rival the compute bill. Bad network design turns billions of dollars of GPUs into idle metal. So the stack deepens: silicon, memory, packaging, cooling, networking, power. Every layer adds cost. Every layer adds constraint.

The costs add up fast. A GPU can cost tens of thousands of dollars, but the chip alone is useless. It needs a server, networking, power distribution, cooling, and the building itself. All-in, the deployed cost per

GPU runs two or three times the sticker price. At the scale of modern deployments, that becomes staggering—a major cluster can cost billions in hardware before anyone pours concrete. On top of these one-time capital expenditures are the ongoing operating costs: electricity, cooling, maintenance, staffing.

But the economic force that shapes behavior most is obsolescence. AI hardware loses value fast. Not because it breaks, but because it gets outclassed. The next generation is always coming. The performance gap can be large enough to make last year's hardware uncompetitive for the most demanding work. Operators push for high utilization. A GPU sitting idle is still depreciating. This depreciation risk shapes a fundamental choice: do you build your own infrastructure or rent from a cloud? Owning can be cheaper if you keep the hardware busy. Renting is more flexible if your demand spikes and drops, or if you do not want to gamble on what the next generation does to your balance sheet. Most big players do both, because uncertainty itself has value—the ability to shift between owned and rented compute is a hedge.

Capital intensity pulls finance into the story. Companies that cannot fund billion-dollar deployments from cash have developed creative structures: debt backed by long-term contracts, sale-leasebacks that turn buildings into liquid capital, investors buying infrastructure as an asset class. To investors, it looks like a gold rush. But the risks are real: rapid obsolescence, tenant concentration, and power constraints that can freeze growth even when money is ready.

One private equity partner, *David*, spends months in 2024 trying to understand chip supply. Everyone is obsessed with chips—waiting lists, secondary markets, allocation tiers. Then he hears something that does not fit: chips sitting idle. Not because no one wants them, but because the power infrastructure is not ready to run them. "Chips are a stock problem," he later says. "You can accumulate them. Power is a flow problem. You use it or you lose it." Same lesson, again and again. Silicon is bound to electricity.

2.6 Where the Chips Go

At the far end of this story, deployments keep growing. GPU clouds sprout across the country. Big tech announces campuses that draw power at the scale of a power plant. Some projects are real; some are aspirational. All of them run into the same shared constraints: chip production, cooling hardware, skilled labor, and, increasingly, grid access.

By spring 2026, the first trucks carrying computing equipment will arrive at the construction site in Saline Township. Data halls may still be unfinished, but the infrastructure to receive the chips must already be in place: cooling loops, electrical substations, fiber connections. The chips represent the endpoint of everything we have described—fabrication and packaging and integration, all converging on what used to be soybean fields. For residents, the chips remain abstract: boxes arriving on trucks, disappearing into buildings where visitors are not allowed. What they see is the transformation around the chips—cleared land, steel and concrete, power lines, substations, debates about water and taxes and jobs.

Silicon is invisible when it works. Infrastructure is not. A chip sitting in a warehouse is idle capital, depreciating by the day. Value comes only when installed in a functioning data center. So we follow the chips into the buildings that house them. Next: inside the data center.

CHAPTER THREE

The Data Center

The construction site stretches across five hundred and seventy-five acres of Michigan farmland. Three massive data halls rise against a gray winter sky, skeletal frames of steel and concrete on what used to be soybeans. Mud cakes everything—trucks, boots, the sides of portable offices. The air smells of diesel and turned earth.

Steve, the site superintendent, stands beside a pickup truck and gestures toward the nearest structure. "That's Data Hall One," he says. "When it's finished, those walls will hold more computing power than existed in the entire world twenty years ago." He has spent three decades building things most people never notice. He started framing houses, moved into commercial construction, and found his way to data centers when a contractor needed someone who could read electrical prints. Now he oversees crews laying the foundation for what the state calls a historic investment.

"Back then," he says, scrolling through photos on his phone, "we were building server rooms in office parks. Twenty kilowatts per rack, maybe thirty—enough to power ten or fifteen homes from each rack. Air conditioning could handle it. A good HVAC crew was all you needed." He pauses on a 2019 photograph: rows of servers with blue lights stretching into the distance like a cathedral. "Beautiful place. State of the art. Totally obsolete for what we're building now."

CHAPTER 3. THE DATA CENTER

Liquid cooling piped to every rack. Power densities that would have sounded ridiculous five years ago. Construction schedules that keep crews working long shifts, week after week, through Michigan winters. "Most folks drive past data centers every day and never think about what's inside," *Steve* tells us. "They see a windowless building. They don't see the engineering that keeps these machines running around the clock." He points toward a trench along the perimeter where workers lay copper and coolant piping. "Cooling. The chips generate so much heat, we can't use air anymore. We pump liquid directly onto the processors. The whole building is basically one giant radiator."

This is what "the cloud" looks like when it becomes a physical address.

3.1 The Box

At its simplest, a data center is a building designed to house computers. But that definition misses the point. A modern data center is a carefully orchestrated system where power, cooling, networking, and physical security work together to keep servers running continuously. If any one of those systems fails badly enough, the building stops being a data center and becomes an expensive box.

Data centers are older than most people realize. In the 1950s and 1960s, they were rooms built around computers that ran hot and failed often, requiring constant attention from operators in white coats. Then came the internet boom, which professionalized the industry: colocation facilities offered space, power, and connectivity to companies that did not want to build their own infrastructure. Cloud computing pushed scale and standardization further, turning data centers into factories that could be replicated across the globe. Now AI changes the building again, forcing a density that transforms it into something new—not just a warehouse for servers, but a factory for heat removal.

3.1. THE BOX

To understand the building, start with the rack. Everything else is designed around this standardized frame that holds servers, switches, and cables. Racks come in neat rows because order is not aesthetic here; it is operational. When something fails at 3 a.m., a technician needs to find it, replace it, and bring it back without guessing. Chaos costs money and time.

Traditional server racks were built for air: bring cold air in, move hot air out. Contain the hot aisle so it does not mix with the cold. Move enough air, and the servers keep running. But AI pushes past those limits. Modern AI racks can draw well over a hundred kilowatts—enough to power fifty homes from a single rack—and at that level, the amount of air you would need to move becomes absurd. Fans would eat power and scream at dangerous levels. Physics sets hard limits on air cooling. So racks become plumbing endpoints, each one needing a way to deliver power safely, remove heat quickly, and move data at high speed. The building supplies all three.

Inside the rack, inside the server, we finally see the machine. An AI server is not a simple web server. Larger. Heavier. CPUs are present, but accelerators are the point—GPUs or other specialized processors designed for parallel math. High-bandwidth memory keeps processors fed with data. Fast interconnects let processors share work without waiting. High-speed network cards connect the server to the rest of the cluster.

All of that needs power. All of that turns power into heat. All of that must be kept inside safe operating limits, every hour of every day. So we move outward from the rack to the building systems that keep it operational.

3.2 Power Never Stops

Power is the lifeblood. Before a data center can do anything else, it must have electricity—and the reason power infrastructure accounts for so much of a facility's cost and complexity is simple: even a brief interruption can corrupt work that has been running for weeks. Three months of training, destroyed by seconds of power loss.

Electricity arrives from the utility at high voltage. For Saline Township, that means a connection into the local utility's transmission network. Engineers step it down through a ladder of voltages until it reaches levels the equipment can use. Then it has to be protected. But the grid is not perfect: storms knock out lines, substations fail, and operators sometimes shed load in emergencies. A facility that went dark every time the grid hiccuped would be useless.

So data centers build redundancy. First layer: uninterruptible power supplies—enormous batteries that take over instantly if utility power drops. They do not run the facility for long, just minutes. Second layer: generators. Rows and rows of them, each the size of a shipping container, ready to convert diesel fuel into electricity on a moment's notice. When utility power fails, batteries keep servers running while generators start. Within a minute or two, the facility can run off-grid, as long as fuel holds out.

Engineers describe redundancy levels with a simple notation: N is the capacity you need, N+1 means one extra path, and 2N means everything is fully duplicated so either half can run the entire building. Full duplication is expensive, but it is the only way to deliver what users assume: that the service is simply there. When we type a question at 2:47 a.m., we do not hope the wind is calm over a substation in Michigan. The whole point of a data center is that we never have to think about it.

3.3 Heat Never Sleeps

If power is the lifeblood, cooling is the problem you never stop solving.

3.3. HEAT NEVER SLEEPS

"You want to know the secret of this business?" *Steve* says, watching his crew lower a coolant distribution unit into place. "It's not computers. It's plumbing." He laughs, but he is not joking. Computing generates heat because physics says it must. When electrons flow through circuits, they encounter resistance. Energy turns into thermal energy. In a data center, essentially all electrical energy entering the building ends up as heat that must be removed.

Steve learned this the hard way years ago, when a cooling system failed and a server room hit dangerous temperatures in minutes. "Smelled like burning plastic and money," he says. "After that, I started paying attention in the mechanical briefings." For decades, air cooling was enough: push cold air through racks, pull hot air out, repeat. But above a certain density, air becomes impractical. Modern AI facilities shift to liquid cooling.

Direct-to-chip cooling is the most common approach. Cold plates mount directly onto processors: precisely machined blocks of metal with internal channels. Coolant flows through, absorbing heat at the source. Warm coolant goes to a heat exchanger, dumping its heat into the building's chilled water loop and eventually out to the environment. Liquid cooling can be efficient. It can lower overhead and improve PUE—Power Usage Effectiveness—the ratio between total facility power and the power used by the IT equipment itself. A PUE of 1.5 means that for every unit of IT energy, the facility uses half a unit more for cooling and infrastructure—about a third of total energy is overhead. A PUE of 1.2 is excellent. But it also adds complexity. Now you are running pressurized fluid through a building and into racks full of expensive electronics. Valves, pumps, filters, leak detection, quick-disconnect couplings designed so technicians can swap a server without draining an entire system.

In Saline, there will be hundreds of coolant distribution units, each one a link between the building's chilled water and the server-level loops.

CHAPTER 3. THE DATA CENTER

Every connection point has to be reliable. A single leak can destroy hardware quickly. At this point, the data center is not a box with computers. It is a thermal plant.

Cooling also drags in a different local resource: water. Many facilities use evaporative cooling towers, where water absorbs heat from the facility and then evaporates, carrying that heat into the atmosphere. Thermodynamically efficient, but thirsty. At large scale, millions of gallons.

In water-stressed places, this becomes political. Communities ask why data centers should receive water allocations that residents and farmers might need. Liquid cooling can reduce total cooling load and cut water use, but it does not erase the issue. Some operators go further, rejecting heat through dry coolers—giant radiators that use no water. These work well when the outside air is cool enough to absorb the heat, but struggle in hot climates where the temperature difference narrows.

Michigan's climate offers the Saline facility some relief. More cool months mean dry cooling can do more work. But summer peaks still push systems to their limits, and local questions do not go away. Water is not only a utility input; it is a neighbor's well, a township's groundwater, a community's sense of risk. In Saline, those concerns show up in consent agreements and monitoring provisions. Cooling is not just engineering: it is politics.

If water is the fight, heat is the opportunity. Liquid cooling captures heat at temperatures that can be useful. Some European data centers sell waste heat to district heating networks. In Michigan, a smaller project in Lansing proposes using data center heat to replace an old steam system with a hot water network. At small scale, you can imagine a clean loop: computing warms water, water heats buildings. At gigawatt scale, the math turns against you. A hyperscale AI facility produces enough waste heat to warm a city, and most American cities lack district heating infrastructure that could absorb it. So the heat dissipates into the atmosphere—not because operators are careless, but because physics wins.

3.4 Wires and Walls

A data center without networking is just a heated warehouse. External connectivity comes through multiple fiber routes, so if one cable is cut by construction or damaged by a storm, traffic reroutes automatically. But the harder challenge is inside the building. Training a large model requires thousands of processors exchanging data constantly. When processors spend time waiting for data instead of computing, expensive hardware sits idle.

Traditional networking can struggle with these traffic patterns. When thousands of processors need to talk to each other constantly, ordinary networks become bottlenecks. So AI facilities use specialized approaches: networking hardware designed to move data between processors as fast as possible, with minimal delay. Think of it as building a private highway system inside the building, where every processor can reach every other processor without waiting in traffic. Different builders make different choices, but the goal is the same: eliminate the waiting. An AI data center is not just "more servers." It is a different kind of system.

All of this infrastructure needs protection. Security starts at the property boundary: fences, cameras, a gatehouse where guards verify credentials before anyone enters. Inside, the building divides into zones with progressively tighter access. Employees badge through doors, scan biometrics, and pass through airlocks where one door must close before the next opens. The precautions might seem excessive until you consider what is inside: hardware worth hundreds of millions, models and services worth billions, and training runs that a single disruption could corrupt. National security concerns add another layer.

CHAPTER 3. THE DATA CENTER

Operational security requires its own boundaries. People who manage the building should not automatically have access to the data. In Saline, three parties share the facility: Related Digital builds and owns it, Microsoft operates the computing environment, and OpenAI runs AI workloads as a tenant. Each has access only to what its role requires, so no single party controls everything.

3.5 Failure Modes

For all the redundancy, things still go wrong. Power failures are the classic nightmare. When utility power disappears, battery backup systems take over within milliseconds. Generators start. Load transfers. That handoff is the moment of greatest vulnerability. Batteries fail? Servers drop instantly. Generators fail to start? Batteries drain. Transfer produces voltage anomalies? Sensitive equipment malfunctions.

Cooling failures can be just as unforgiving. Without heat removal, temperatures climb fast. In high-density systems, minutes matter. Human error adds another risk: pulling the wrong cable, flipping the wrong breaker, performing a maintenance procedure out of sequence. Data centers are full of systems that usually work—and people who sometimes make mistakes.

Reliability is not just engineering. It is culture. The best facilities treat even minor failures as lessons, documenting everything to prevent recurrence.

All of this infrastructure requires people. A large data center employs hundreds: facilities engineers, security, network technicians, operations staff monitoring thousands of sensors around the clock. AI facilities expand the skill set further. Liquid cooling means technicians must understand coolant chemistry, pressure and flow, and the difference between a minor anomaly and a leak that could take down a row. The work is physical, the environment demanding. Safety training is extensive because electrical faults can be lethal and high-energy systems can fail violently.

We can say "the data center runs itself" only because people make it feel that way. And beyond those inside the facility, millions of workers now depend on what runs here. Software developers use coding assistants. Researchers rent GPU time. Customer service teams rely on automated agents. The systems that serve hundreds of millions of people are housed in facilities run by a few hundred.

3.6 Building at Scale

Building a data center also means building a supply chain. Transformers from one supplier. Switchgear from another. Generators from a third. Servers assembled in one place from components made in several others. Cooling systems, pumps, heat exchangers, specialty fittings from yet more manufacturers. Lead times stretch from months to years, and delays cascade. The AI boom adds new bottlenecks: not only chips, but cooling components, specialty couplings, and heavy electrical equipment required to move large amounts of power.

For a project like Saline, supply chain management starts years before the first shovel hits the ground. Teams order major electrical equipment far in advance. They place server orders before completing final designs, just to secure allocation. Even delivery is a project. The biggest transformers weigh hundreds of tons and require specialized transport. Roads must be assessed, permits secured, deliveries coordinated with utilities and local authorities. Anyone who says "data centers are just warehouses" has never tried to deliver a hundred-ton transformer to a rural township on a narrow road.

So why do data centers keep getting larger? Fixed costs. A facility needs switchgear whether it serves a hundred megawatts or a thousand.

CHAPTER 3. THE DATA CENTER

Security perimeters cost nearly the same for small sites as large ones. Engineering teams must be hired regardless of scale. Spreading fixed costs across more capacity lowers cost per megawatt. Campuses grow because scale is an economic necessity.

AI pushes the math harder, because the hardware is dense and expensive. The economics work only if the facility stays utilized. An empty data center still has fixed costs. A partially utilized campus still pays for the same security and much of the same power and cooling infrastructure. Developers want long-term commitments before they build. Tenants sign deals before the first rack is installed. Less speculative real estate, more industrial planning.

What makes AI different is how these systems fuse together. In older facilities, power, cooling, and networking were important, but the margin for error was wider. AI systems operate at higher power densities, narrowing that margin. Cooling failures that once gave operators half an hour can become emergencies in minutes. Training runs can be corrupted by interruptions that a web server would shrug off. As reliability requirements rise, so do the requirements for redundancy.

Silicon described in the last chapter does not matter without the building. And the building does not matter without the silicon. This integration is one reason hyperscalers hold such asymmetric power right now. They have spent years assembling the physical assets (chip allocations, land near transmission capacity, facilities) and the relationships needed to operate them. In a market defined by lead times measured in years, being early is an advantage money alone cannot buy.

Buildings being constructed today will operate for decades. The equipment inside will not. Each hardware generation brings higher densities, new cooling demands, different power requirements. Builders design for this uncertainty, leaving room for upgrades: extra cooling capacity, scalable power distribution, fiber sized for future bandwidth. *Steve* lives with this paradox daily.

3.7 Inside the Hall

Most people will never see inside an operational AI data center. The buildings are intentionally anonymous. Step through the airlock—one door closes before the other opens—and the transition is immediate.

First impression: sound. A constant drone of fans and mechanical systems, not deafening but impossible to ignore. Second impression: order. Racks in precise rows. Overhead cable trays carry neatly bundled fiber, color-coded by function. Nothing here is accidental. Racks bristle with cooling connections. LEDs blink in patterns meaningful only to trained eyes. Coolant pumps hum. Power distribution gear feeds electricity through cables thicker than a person's arm.

None of this is visible to the user. Researchers see a software interface. Students see a text box. Professionals see a tool that makes work go faster. All of that physical reality exists only to make the interface feel simple: power, cooling, networks, security, and people.

We have followed the token backward through inference, through silicon, into the building where mathematics becomes physical. But buildings are passive. They house equipment. They do not power it. The gigawatt appetite must be fed continuously, around the clock, without interruption. Next we confront the constraint that shapes everything else: power.

CHAPTER FOUR

The Power Constraint

Sarah arrives at PJM Interconnection headquarters in Audubon, Pennsylvania, before sunrise on a Tuesday in December 2024. The parking lot sits nearly empty. The building is quiet and anonymous—a low-slung office complex surrounded by bare trees and frozen grass. Nothing about it looks like a nerve center. But inside, PJM coordinates the wholesale electricity grid across thirteen states and Washington, D.C., managing power flow for tens of millions of people. A single mistake here can ripple into catastrophe.

Sarah is a senior grid operations analyst. She starts early because the interconnection queue waits for no one. The queue is the formal process through which new power plants and large electrical loads—major electricity consumers like data centers—get permission to connect to the grid. Utilities must study whether the grid can handle each new load before approving connections, and the resulting wait can stretch for years. When she first joined PJM, a big request might be fifty megawatts. Unusual, but manageable. Now her dashboard shows something else entirely: data center demand piling up faster than her team can process the studies. Dozens of gigawatts under contract or active negotiation.

She scrolls through the morning's new applications. Three more gigawatt-scale requests: one from Michigan, two from Virginia. She clicks the Michigan file. Related Digital. Saline Township. Something

CHAPTER 4. THE POWER CONSTRAINT

called Stargate. The number in the capacity field makes her blink: 1.4 gigawatts. That could be a substantial share of a major utility's peak load. *Sarah* whistles softly to no one in particular. Welcome to another Tuesday.

4.1 WHAT BINDS

Every constraint shapes what can be built. In software, the constraint is often time—ship before a competitor ships. In chip manufacturing, physics sets the limit—how small can you make a transistor, how tight can you pack heat. For AI infrastructure, one constraint dominates all others. Power. Not computing power. Electrical power. The raw flow of electrons measured in megawatts and gigawatts.

Power determines nearly everything else: where data centers can be built, how quickly they can come online, who can afford to wait long enough to build them. We hear endless talk about GPUs and funding rounds. Those matter. But the thing that makes or breaks these projects is simpler: how many electrons can you reliably obtain, year after year, at prices that make the business work.

Understanding the power constraint requires a quick primer on electricity. Electricity is the flow of electrons through wires. Voltage is the pressure pushing them. Current is how many flow per second. Power, measured in watts, is the rate of energy transfer. Think of water in pipes: voltage is pressure, current is flow rate, wattage is how much work the flow can do.

Each prefix is a thousandfold jump. A kilowatt runs a toaster. A megawatt runs a small factory. A gigawatt is a power plant running flat out, just to keep one customer happy. One more distinction: capacity versus consumption. Capacity is peak demand—how much power a facility can draw at any moment. Consumption is energy over time—how much it uses across hours, days, and months.

Utilities plan around peak demand because the grid must balance instantaneously. Every watt consumed must be generated at that exact moment. Storage exists, but at grid scale it remains limited. So the hard question is not "How much energy will this data center use over a year?" The hard question: can we deliver that full gigawatt-plus, on the hottest day of August, at 3 p.m., when the region is already straining? Data centers care about capacity. And capacity explains why the map of AI infrastructure looks the way it does.

4.2 Why Not Cities

If you have followed technology news over the past decade, you might picture data centers as sleek facilities in metropolitan areas. Colocation companies built businesses on being near big internet exchange points. Fifty or a hundred megawatts once seemed huge. The AI era renders that model obsolete. New campuses want five hundred megawatts—enough to power a city of four hundred thousand people. A thousand megawatts. Sometimes more.

Loads like that cannot locate in cities. Not because cities lack connectivity, but because city electrical infrastructure cannot deliver gigawatt loads. Urban distribution networks—wires under streets, transformers on poles—serve thousands of customers at modest power levels. An urban substation might handle a couple hundred megawatts for an entire neighborhood. A gigawatt-scale facility must bypass local distribution and connect directly to high-voltage transmission lines.

Upgrading urban infrastructure to carry that kind of load is possible in theory. Brutal in practice. It means tearing up streets, negotiating easements, relocating existing infrastructure, managing years of construction in dense areas. Expensive. Slow. Data center developers do not wait five to ten years for an urban retrofit. They go where high-voltage transmission lines already run. And transmission corridors run through rural land, because rights-of-way are easier there and the lines were built to connect power plants to load centers.

The result: some of the most advanced computing facilities on Earth end up on farmland, not because developers love farmland, but because electrons already flow there. AI infrastructure follows transmission maps, not talent maps. Northern Virginia became the world's first data center megacluster partly by accident. Substations and transmission lines built to serve government facilities created spare capacity. Early internet companies found it. The cluster grew. Once it existed, it pulled more development toward itself. But even Northern Virginia has limits. Eventually saturation pushes outward into adjacent counties—wherever the grid still has room.

That pattern repeats everywhere. Saline Township sits fifteen miles from Ann Arbor and twenty from Detroit, but the project was chosen for neither the university nor the workforce. The project landed in Saline because of the transmission corridor that crosses it.

Being near transmission is not enough. Connecting to the grid requires navigating the interconnection queue: a regulatory and engineering process that has become the binding constraint on new development. The queue has exploded. On the generation side, endless proposed power plants wait for permission to connect; many will never be built. Utilities see a parallel surge on the load side: large customer requests dominated by data centers. Wait times have stretched from two years to five, eight, even ten. For a technology industry accustomed to six-month product cycles, this pace feels incomprehensible.

The queue is not just slow; it is also expensive. Projects pay for grid studies. If those studies reveal needed upgrades, projects can be asked to contribute—new substations, new lines, bigger transformers. At large scale, those costs reach into the hundreds of millions. A small company cannot absorb years of delay plus huge upgrade costs. Only large, patient capital can survive the queue. So the interconnection process concentrates AI infrastructure among the biggest players.

4.3 POWER FIRST

Inside the industry, a blunt principle emerges from hard experience: power first. Available electrical capacity and delivery timelines outrank everything else. Site selection has been inverted. When a hyperscaler plans a new AI campus, it begins with utilities, not real estate. Teams identify service territories with available capacity, assess which utilities have realistic timelines, look at transmission topology. Only then do they examine specific parcels. Highway access, fiber, and labor markets matter too, but only after power is secured.

The result is locations that defy traditional corporate logic: small towns in Kansas, Nebraska, and rural Michigan, chosen not for talent pools or transportation hubs but because transmission lines cross their land. Power first also changes the relationship with utilities. A decade ago, power was something you assumed you could buy. Now you negotiate for it years in advance. Developers cultivate utility executives, structure long-term contracts, and seek regulatory approval. Utilities that can deliver power at scale draw investment; those that cannot get passed over. This explains why places like Virginia and Texas have become major hubs.

GPU shortages dominate headlines, but investors who look closely find a different constraint. One of them, *David*, learns this before many of his peers. He reviews a deal that has everything: anchor tenant commitment, chip allocation locked, construction team ready. What it lacks is grid access. The queue shows a seven-year wait. The deal dies.

CHAPTER 4. THE POWER CONSTRAINT

He starts calling utility executives and learns how the process works. "Power access is the moat," *David* says later. "Chips you can eventually buy. Power takes years." His fund pivots. They prioritize projects with secured grid capacity, or at least positions far enough along in the queue that the timeline is credible. Capital reacts to constraints this way. When a resource binds, everyone starts paying to secure it.

Spending patterns confirm it. The biggest companies are not only signing chip contracts. They pay for substations. Fund transmission upgrades. Look for new generation. Pursue nuclear partnerships that would have sounded like science fiction a decade ago. Those moves make sense only if power is scarcer than chips.

4.4 Who Pays

"Power first" plays out on the ground in predictable ways. Northern Virginia dominates, then hits saturation and pushes outward. Texas attracts projects partly because its interconnection process moves faster. It also tolerates behind-the-meter generation: data centers building their own power on site, often natural gas, to bypass the queue. Michigan is not a classic data center market. Its electric system was built around manufacturing. Yet Saline becomes a contender because the transmission corridor exists, and because the utility is willing to invest.

The power constraint raises a question that moves from engineering to politics: who pays? When a utility builds infrastructure to serve a data center, those costs often enter the rate base—the total capital investment used to calculate everyone's electricity bill. Ratepayers are ordinary electricity customers: families, small businesses, anyone who pays a monthly bill. Should they subsidize infrastructure built primarily for data centers? Utilities argue that large customers bring revenue and spread fixed costs, improving system economics. Critics counter that socializing costs shifts risk onto households, especially when contracts stay opaque.

In some regions, these costs show up in capacity market prices and rate cases. In others, contract terms are redacted, making independent verification impossible. Data centers become political flashpoints for this reason. They are not only private projects. They plug into public systems. And at this scale, they change those systems for everyone.

4.5 Why Saline

We began with a question: why build a seven-billion-dollar data center on Michigan farmland? The answer is power. Saline wins not because of universities or workforce or tax policy, though those matter at the margin. It wins because transmission infrastructure crosses that land with the capacity to serve a gigawatt-scale load. Electrons flow there. Everything else follows.

Once we see that, the rest of the AI infrastructure story opens up. Data centers land on farmland because transmission corridors cross it. Timelines stretch to five or ten years because interconnection studies, approvals, and construction each take time. Utilities become essential partners because no one else can deliver electrons at this scale. The next chapter follows those electrons into the grid itself: the system that moves power from generation to consumption.

CHAPTER FIVE

The Grid

A summer thunderstorm rolls across Northern Virginia at 4:47 p.m. on a Wednesday in July 2025. *Sarah* watches the transmission map on her screen turn yellow, then orange. She has spent more than a decade at PJM Interconnection, the regional operator that balances electricity across thirteen states. Wall-sized displays show power flowing in real time: demand rising and falling, generators ramping up and down. *Sarah* has seen heat waves and polar vortexes. She has never seen anything like the past two years.

Loudoun County spikes. Data Center Alley—the nickname for Northern Virginia's concentration of data centers, the densest cluster in the world—pulls hard. Six gigawatts of operational load concentrate in one county, enough electricity to power four million homes, with more under construction every month. Demand that used to be a rounding error now commands the whole shift's attention when storms roll in. Lightning strikes near a substation. A high-voltage line trips.

For a fraction of a second, millions of dollars of computing equipment teeter between operation and shutdown. Automated systems respond: backup circuits engage, data centers switch to batteries, diesel generators cough to life. The internet stays up. Millions of AI queries continue without interruption. Users never see the spinning cursor, never know how close they came.

Sarah types "non-event" into the incident log. She knows better. Margins are shrinking. Every gigawatt added to the grid reduces the buffer between normal operations and trouble. To understand why data centers rise where they do, we have to understand the system they plug into—and why that system resists fast change.

5.1 Three Grids, Many Rules

Americans talk about "the grid" as if it were one unified machine. It is not. What we call the grid is three weakly connected systems, dozens of operators, and overlapping jurisdictions. Thousands of local decisions over a century built it, none with twenty-first-century computing in mind. The Eastern Interconnection. The Western Interconnection. And Texas, which operates largely as an island through ERCOT, the Electric Reliability Council of Texas. All three run at sixty hertz, but they have limited ability to help each other when trouble hits.

A power plant in California cannot easily send electricity to Chicago. A generator in Arizona cannot rescue New York during a heat wave. Each interconnection is, for practical purposes, its own continent. Within each, utilities carve out service territories: investor-owned corporations, cooperatives, municipal utilities, and federal entities like the Tennessee Valley Authority. All operate under different incentives and constraints.

Regulation is split, too. The Federal Energy Regulatory Commission governs wholesale markets and interstate transmission. State public utility commissions regulate retail rates and approve generation. County governments influence permitting. Environmental agencies can slow or block projects. Building a transmission line across state boundaries can require approvals from multiple bodies, each with its own timeline, each with its own opportunities for opposition. The grid changes slowly. A business that wants power in months runs headlong into a system that moves in years.

5.1. THREE GRIDS, MANY RULES

Over the past few decades, regional transmission organizations and independent system operators—RTOs and ISOs—have reorganized much of the grid. These entities own no power plants, no transmission lines. They run markets, dispatch power, maintain reliability, and manage interconnection: the process of connecting new generators and large loads to the grid. For data centers, RTO and ISO rules are not a footnote. They are the rules of the game, shaping the two things that matter most: electricity price and the timeline for getting it.

Consider three grid operators: PJM in the Mid-Atlantic, ERCOT in Texas, and MISO in the Midwest. PJM is the largest RTO in North America, covering much of the eastern U.S., including Northern Virginia and its dense cluster of data centers. That history creates momentum. It also creates congestion. The interconnection queue has swollen, wait times stretch for years, and the process has shifted from first-come to clustered studies because too many projects now compete for attention. For developers, the old playbook breaks. You cannot just buy land, apply, and wait. The wait is too long. So developers chase alternatives: behind-the-meter power or regions where interconnection moves faster.

Texas works differently. ERCOT operates the Texas grid as an island, and by staying mostly within state boundaries, it avoids some federal oversight. Its market structure favors speed and competition. Interconnection can move faster than in other regions. For data centers, speed is the attraction. The risk is reliability—an island grid cannot import power from neighbors in an emergency. Winter Storm Uri in 2021 demonstrated the risk: when a truly bad event hits, an island grid has fewer backstops.

MISO—the Midcontinent Independent System Operator—covers a huge swath of the country, including Michigan and Saline. For data centers, the appeal is practical: queues less saturated than PJM's, cold winters that reduce cooling costs, available land, and utilities eager for new industrial customers as older manufacturing declines.

5.2 Pipes and Highways

A physical distinction explains the siting map: transmission versus distribution. Transmission is the high-voltage backbone—steel towers marching across the countryside with cables humming at hundreds of thousands of volts. High voltage allows power to move long distances efficiently, with lower losses. Distribution is the lower-voltage final leg to homes and businesses: wooden poles along residential streets, neighborhood transformers, wires sized for a world of small, scattered loads.

A household draws a couple of kilowatts. A grocery store draws tens or hundreds. A neighborhood substation serves a few thousand homes. Now compare that to a modern AI facility, which might want five hundred megawatts or more—the equivalent of hundreds of thousands of homes. Grid operators use a simple analogy: transmission is the main pipe from a reservoir, distribution is the thin plumbing in your house. You cannot fill a swimming pool through a kitchen faucet. The pipes are too small.

Urban distribution cannot serve a gigawatt-scale data center without rebuilding from scratch. So these facilities connect at transmission voltage. They build or expand substations, run dedicated lines to their sites, and pay for studies and upgrades to ensure the new load does not destabilize the grid. All of this takes time and money, and it imposes a geographic constraint: data centers must follow transmission corridors, which often means rural land.

5.3 The Queue

Here is where the queue comes in. The interconnection queue is where data center ambition meets infrastructure reality. Every large load that wants to buy power, every generator that wants to sell power, applies for an interconnection study. That study determines what upgrades are needed, who pays, and when the project can operate. In theory, a rational allocation mechanism. In practice, the bottleneck that shapes the entire buildout.

Why does the queue move so slowly? Several forces collide. The energy transition floods queues with renewable projects, each requiring study. Speculative applications clog the system—developers reserve queue positions at multiple sites to keep options open. Data center load growth adds enormous continuous demand, single facilities drawing as much power as medium-sized cities. And transmission infrastructure has not kept pace. Building new lines can take a decade.

For data center projects, the numerous studies required, from feasibility and system impact to facilities, compound the time requirement. Only after these studies are completed and approved can a project sign an interconnection agreement and begin construction. Construction takes more time still. Even without new transmission lines, the cumulative timeline can stretch to five years or more. For a company racing to monetize AI compute, that mismatch is existential. Developers adapt. That adaptation shapes land use.

5.4 Why Farmland Wins

A paradox sits at the center of data center development. Policy prefers redeveloping brownfields—previously developed land like old mills, coal plants, factories, malls, and parking lots, land already scarred by prior use. Environmentalists worry about consuming greenfields—undeveloped land like farms, forests, and wetlands. And yet data centers keep rising on clean agricultural land. America has thousands of idle brownfields. Developers skip them. Why? Grid topology.

The intuition that "old industrial sites already have power" is often wrong. Those sites had connections sized for twentieth-century loads, and many were decommissioned. Substations were not built for twenty-first-century data center demand. When requirements change, queue positions reset. What used to work at a hundred megawatts does not automatically work at five hundred. Brownfields also come with slower timelines: contamination studies, cleanup, complex title, neighbors with legitimate reasons to fight.

Compare that to farmland near a transmission corridor. Flat land. Clean title. Faster environmental assessment. And the path to power is often shorter because the big lines already run nearby. When a developer faces a five-to-seven-year interconnection timeline and a competitive race, the cornfield wins. Not because developers hate cities. Because the grid makes some sites structurally advantaged.

Yet the environmental footprint extends beyond the data center site. New transmission capacity often means new rights-of-way cutting through forests, wetlands, and undeveloped land. A single high-voltage line can require clearing a corridor a hundred feet wide or more, stretching for dozens of miles. Trees felled. Wetlands crossed with access roads and tower foundations. Habitats fragmented. The data center may sit on former farmland, but the infrastructure that feeds it touches ecosystems that were never plowed. When we count the land consumed by AI infrastructure, the campus footprint is only part of the story.

When the queue becomes intolerable, some developers take a more radical step: they generate their own power on site, behind the meter. A behind-the-meter facility still connects to the grid, but it installs its own generation—usually natural gas turbines—capable of meeting most or all of its load. This gives operators flexibility: they can draw from the grid when electricity is cheap and switch to on-site generation when prices spike, or they can run their own generators continuously and treat the grid as a backup. The appeal is obvious: bypass the queue.

Building a gas plant takes a couple of years; waiting for grid interconnection can take five or more. A developer facing that choice often picks the faster option: power in 2027 with on-site gas generation rather than power in the early 2030s from the grid. The environmental implication is that time pressure favors fossil fuels, even when companies have long-term commitments to renewables. Companies argue that gas now plus renewables later still beats waiting years for cleaner grid power, but critics see it as greenwashing.

This trend strains traditional regulation. The traditional model assumes that utilities generate power and customers buy it. When customers build their own power plants, utilities lose sales but may still need to maintain backup infrastructure. Regulators must decide how to charge for standby service and prevent cost shifting. Rules catch up slowly. The industry moves fast.

The power constraint that drew the Saline project to Michigan farmland raises a practical question: how does the utility actually plan to deliver gigawatt-scale capacity? DTE, Michigan's largest utility, operates within MISO, where queue times are shorter than PJM's but still measured in years. DTE has its own pipeline of requests. Yet the utility claims it can serve Saline's contracted capacity by leaning on existing infrastructure, prioritization, and upgrades financed by the project. Whether that promise holds is the bet.

What we can see clearly is the hidden asset: the transmission corridor. High-voltage lines cross the township, built decades ago to serve the industrial geography of southeast Michigan. Manufacturing has declined. Some capacity is now available. That opportunity does not exist on an unconnected site. A developer looking at Saline does not see soybeans. They see a corridor with voltage, a substation that can be expanded, and a path to power shorter than most alternatives.

5.5 The Cost Shift

The question of who foots the bill is significant. When a utility builds infrastructure to serve a data center, who pays? Some costs flow through

CHAPTER 5. THE GRID

capacity markets and transmission upgrades. Demand growth raises the need for capacity and the cost of maintaining margins. Those costs spread across all customers in a region, even if the data center sits in one county. A homeowner in one state can feel the effects of data center growth in another.

Utilities and industry advocates argue that large load growth can lower average prices by spreading fixed costs across more sales. Sometimes true, but not always. When infrastructure is constrained and expensive, when new capacity must be built rather than merely used, costs can rise before any long-term benefits arrive.

In Michigan, the Saline contracts illustrate the problem. Key provisions are redacted, so regulators approve deals that even the state attorney general cannot fully evaluate. Officials argue that protections exist, but critics counter that those protections cannot be verified. For communities asked to host projects of this scale, that trust gap becomes political fuel.

Recognizing the queue crisis, the Federal Energy Regulatory Commission issued Order 2023, the largest interconnection reform in decades. The reforms aim to make the process less wasteful: clustered studies instead of one-by-one, higher financial commitments to discourage speculative applications, timelines with consequences, better coordination across ISO boundaries.

But Order 2023 does not build transmission lines, solve lengthy permitting fights, or staff every ISO with enough engineers to process thousands of applications. It improves the plumbing without expanding the pipe. For data centers, it may turn an eight-year wait into a six-year wait, but for companies that need power in eighteen months, neither timeline works. So the industry keeps adapting: behind-the-meter generation, geographic diversification, and deals that finance their own infrastructure in exchange for priority.

5.6 Saturation

These constraints shape where data centers land. Data centers cluster where grid capacity exists, saturate those areas, then push outward to regions like Texas, where processes move faster and developers accept reliability risk, or the Midwest, where queues are less congested and utilities are hungry for large customers. Rural counties near transmission infrastructure find themselves facing proposals they never expected, weighing tax revenue against a transformation that feels abrupt and unwanted.

For rural communities like Saline Township, these changes arrive at enormous scale. A township of a few thousand people faces a proposal for a gigawatt-scale facility. Investment exceeds local budgets by orders of magnitude. Decisions are shaped by actors operating at state, national, and global scales. The community can resist. It cannot rewrite the grid.

Looking forward, these constraints are not temporary inconveniences. They reflect decades of underinvestment in transmission, regulatory fragmentation, and a planning system designed for incremental growth. The data center industry needs power measured in gigawatts, on timelines measured in months, from a system designed to deliver kilowatts on timelines measured in years. Something has to give.

Either the grid changes, or demand finds ways around it: private generation, locating in less congested regions, new compromises. All of this raises the next question: where does the power come from? The grid determines where data centers can connect. Generation determines the carbon, the fuel, and the environmental footprint of what they do. In the next chapter, we follow the electricity upstream, into the power plants.

CHAPTER SIX

The Generation

The control room at Three Mile Island Unit 1 has been dark for nearly a decade when *Elena* walks back in. In September 2019, the reactor shut down after cheap natural gas undercut its economics. The plant went quiet, operators moved on, and the industry, already scarred by the 1979 accident at the neighboring unit, looked like it was shrinking again. Now, on a morning in the late 2020s, *Elena* stands before the same control panels where she spent years of her career. She watches technicians perform a startup sequence—this time, for Microsoft.

The irony is hard to miss. Three Mile Island, the name synonymous with nuclear fear, is coming back to life to power artificial intelligence. "Criticality in fifteen minutes," the shift supervisor announces. *Elena* watches neutron flux rise as control rods withdraw, bringing the reactor back after years of dormancy.

Her mind drifts to how they got here: a restart announcement in 2024, a long-term power purchase agreement, four years of regulatory review and equipment work. Then she recalls the competing announcement from the week before—a massive natural gas project planned specifically for AI data centers. A comparable gas plant would have been running in half the time. AI demand does not wait for either.

When the reactor goes critical, the first electrons flow toward Microsoft's data centers. *Elena* lets herself feel satisfaction, but she also knows she is watching an exception. Across America, most of the power feeding AI still comes from burning fossil fuels. And it will for years to come.

6.1 THE MIX

Where does data center electricity come from? Electricity is not abstract—it comes from turbines and boilers and reactors and spinning blades, from a grid running on a particular mix of fuels. As of 2025, natural gas dominates at about forty percent of American electricity, followed by wind, solar, and hydro at twenty-four percent, nuclear at seventeen percent, and coal at sixteen percent. These are national averages; any given region can look very different.

Region matters for more than interconnection speed. It determines carbon intensity—whether a "clean power" claim holds up in practice or dissolves into paperwork. But one concept matters even more than the average mix: marginal generation. When a new load turns on, when a data center draws hundreds of megawatts continuously, which power plant ramps up to meet that demand? Usually a gas plant. Even as the grid grows cleaner on average, incremental power for new demand is often fossil. Natural gas is the thread that runs through the story of data center power.

6.2 WHY GAS

Natural gas becomes the default fuel for new data center power because it fits the AI timeline. Gas turbines deploy on two- or three-year horizons, not decade-long ones. Proven technology. Reliable, dispatchable baseload power—electricity that runs twenty-four hours a day, seven days a week, regardless of weather. Unlike solar, which produces only during daylight, or wind, which depends on conditions, baseload power shows up at three in the morning when the wind is calm and the sun is down. And gas plants can be built behind the meter.

The biggest announced data center projects almost all rely on natural gas.

6.3 Gas Costs

Gas does not arrive without consequences. Burning it emits carbon dioxide—Goldman Sachs estimates that data center expansion will add two hundred million tons of annual carbon emissions by 2030. The Washington Post reported that by the mid-2030s, data center emissions could equal those of New York City, Chicago, and Houston combined. Extracting, processing, and transporting that gas also leaks methane, which traps far more heat than CO_2 in the near term. With enough leakage, gas loses much of its climate advantage over coal.

Carbon is not the only impact. Gas plants emit nitrogen oxides and other pollutants, add noise, and spark fights over new pipelines and compressor stations. Behind-the-meter construction can feel to communities like the project is dodging normal checks. The environmental impact of data centers comes not only from gas production and emissions, but from water consumption as well.

In Arizona, a farmer like *Roberto* watches wells drop and wonders how many new industrial customers his aquifer can handle. He grows alfalfa outside Phoenix. He knows water politics the way most people know traffic: a constant, background condition of life. At the edge of his fields, the canal runs lower than it used to. Restrictions tighten. Neighbors drill deeper. Everyone tells the same story with different dates: "When my father farmed, the water was higher."

Then the data center proposals arrive. The numbers get strange. A building that looks like a warehouse, with no smokestack, can still demand huge volumes of water if it relies on evaporative cooling. Water turns into vapor and disappears into the desert air, carrying heat away. It works. It also consumes. *Roberto* goes to a meeting and hears the word "cooling" used like a footnote. Engineers describe their systems with confidence. He realizes the promise being made is not "we will not use water." It is "trust us." A city council votes down one project. Another community approves a different one. Headlines come and go. The aquifer stays. The wells keep dropping. Trade-offs keep happening, often invisibly.

6.4 The Accounting Trick

How can companies building gas-fired facilities also claim renewable progress? The answer lives in corporate carbon accounting, which divides emissions into one of three scope categories. Scope 1: direct emissions from sources you own. Scope 2: emissions from electricity you purchase. Scope 3: everything else—supply chains, manufacturing, logistics, upstream and downstream. Most corporate renewable energy claims focus on Scope 2, and most Scope 2 claims rely on certificates.

Renewable energy certificates—RECs—separate "the electron" from "the claim." When a wind farm generates a megawatt-hour, it creates two things: a physical electron flowing into the grid and a certificate saying renewable generation happened. Those can be sold separately. A data center draws electricity from a grid heavy on natural gas, buys RECs from a wind farm in another state, and claims "100 percent renewable" in a sustainability report. The electrons are no cleaner. The accounting looks green.

This practice is called annual matching: match your annual consumption with an annual purchase of certificates. The practice is both allowed and common. But annual matching does not ensure renewable generation meets consumption when consumption happens. A data center draws power at midnight. Solar generates at noon. A certificate pretends those are interchangeable. The atmosphere does not.

Some companies pursue a stricter goal: hourly matching, sometimes called 24/7 carbon-free energy. Align consumption and carbon-free generation hour by hour, in each region where you operate. It requires storage, or a firm carbon-free source that runs day and night. Hourly matching forces you to confront the clock. Buying solar is easy. Buying solar-plus-storage that can follow demand hour by hour? Much harder. Even storage has limits: batteries can cover hours, but covering days of low wind or long winter nights requires much more.

Companies pursuing 24/7 carbon-free power end up building portfolios, not slogans: renewables, storage, transmission, and some kind of firm clean supply. Here the clean-energy story stops being accounting and starts being engineering. Which brings us back to nuclear.

6.5 THE NUCLEAR BET

We also hear a different defense from the industry: hyperscale data centers are more efficient than scattered enterprise servers. This is often true, but it does not settle the climate question. Efficiency per computation is not the same as total emissions. When compute becomes cheaper, people use more of it. Models get deployed in more products. Agents spin up more background work. Demand expands into places it never would have reached at higher prices. The power plant that serves that extra demand is usually not a new wind farm that happened to come online at the right hour. It is the marginal generator—often a gas plant.

Nuclear energy attracts hyperscalers because it does something renewables struggle to do without enormous storage: it runs steadily. A nuclear plant can operate near full output most of the time, and it is carbon-free at the point of generation. It gives a data center operator what the AI business model craves: predictable, always-on power.

After decades in which nuclear seemed politically untouchable, the technology industry has started making deals—and the timelines depend on which kind.

The fastest path is securing output from reactors already running. Constellation has signed deals with Microsoft and Meta tied to existing plants. These deliver carbon-free electrons immediately but add no new capacity to the grid.

CHAPTER 6. THE GENERATION

Restarts are slower. Three Mile Island Unit 1, the headline example, was announced in late 2024 with a target of 2028—roughly four years of regulatory review and equipment work. Palisades in Michigan and Duane Arnold in Iowa follow similar three-to-five-year timelines. Only a handful of shuttered American reactors are viable candidates.

New construction is slower still. Small modular reactors promise factory-built units and faster schedules than the custom megaprojects that gave nuclear its reputation for delay. But new reactors carry first-of-a-kind risk. Vogtle Units 3 and 4 in Georgia, the most recent large American nuclear project, were supposed to cost fourteen billion dollars; they cost more than thirty-five billion. They were supposed to start operating in 2016 and 2017; they started in 2023 and 2024. The delays bankrupted Westinghouse, the lead contractor. SMR developers aim for first commercial units around 2030—optimistically.

Nuclear remains the best candidate for firm, carbon-free power at scale. Restarts could deliver before the decade is out. New designs are a bet on the next one. Until either arrives, gas fills the gap—and the risk is that the bridge becomes the road.

Geothermal energy deserves a mention. Traditional geothermal works only where hot rock sits close to the surface, limiting it to a few regions. Enhanced geothermal aims to expand that map by drilling deeper and creating artificial reservoirs, and if it scales, it could offer something rare: firm, carbon-free power with a small land footprint. But an early enhanced geothermal plant can power a small slice of a data center, not a gigawatt campus. The technology may matter in the future, but it cannot help with the power demands of today.

6.6 Renewables

If gas dominates the near term and nuclear dominates long-term planning, where do renewables fit? They matter. Costs keep falling. Wind and solar are often the cheapest new sources of electricity in many regions. Data center operators sign power purchase agreements, locking in long-term prices and funding new buildout. Amazon and others have financed huge amounts of new renewable capacity through these contracts.

The challenge is intermittency. Solar produces nothing at night. Wind depends on weather. A data center needing hundreds of megawatts continuously cannot rely solely on sources that fluctuate between zero and more than their rated output. Storage helps, and storage costs keep falling. But storage at the scale needed for a large always-on facility remains expensive and limited by supply chains.

Renewables also depend on transmission. The best wind and solar resources are not always near the load, and the grid is already constrained in many places that want to grow fastest. Even when a company signs a power purchase agreement, physical delivery can be bottlenecked by lines that do not exist yet. That gap between contract and delivery is part of why the clean-energy transition feels slower than the press releases suggest. Hardware is getting cheaper; siting fights are not.

Permits slow projects. Local opposition can stall lines for years. Transformers and switchgear take time. Every delay pushes operators back toward what they can control: gas turbines on their own property. Most renewable strategies look like this: buy renewables to offset annual consumption, stay connected to a grid that provides backup and firm power, use gas—directly or indirectly—to fill the gaps. Over time, emissions may fall. They do not disappear on the timeline the AI industry wants.

6.7 Saline's Power

The Saline facility faces these questions directly. The project plans to consume one point four gigawatts—a huge new load for a utility like DTE. Where will that power come from? DTE's current mix includes a nuclear plant, some renewables, and a lot of fossil capacity. Michigan law ties tax incentives for data centers to "clean energy" requirements. The political deal is clear: incentives in exchange for cleaner power.

But definitions matter. Enforcement mechanisms matter. A project can satisfy a clean energy requirement by buying RECs on an annual basis—checking a box without changing the physical electrons flowing into the facility. On paper, the bar looks high. "Clean" sounds like a promise about the actual power plants serving the load. In practice, "clean" can mean nuclear output already on the system, wind and solar built somewhere else, or certificates bought in a market. Enforcement can be strict, or it can settle for annual matching.

If regulators treat the requirement as physical, hour-by-hour power, the project needs new firm carbon-free supply—exactly what is hardest to build on the AI timeline. If regulators treat it as accounting, the requirement becomes a procurement exercise, not a grid transformation. Practical reality: much of Saline's power will come from natural gas, because that is what can meet marginal demand at this scale, on this timeline. Over time, DTE could add more renewables, invest in storage, pursue firmer carbon-free options. Those changes operate on five-to-ten-year horizons. The data center begins consuming power in 2027.

The pattern repeats across the industry. Press releases emphasize clean commitments. Critics point to gas plants. Both capture something true. The gap between them is the politics of the AI buildout.

What does all of this mean for emissions? Projections vary, but direction is clear: if data center demand grows quickly and marginal power is gas, emissions rise. That outcome is not destiny. Storage could get cheap enough to matter at scale. Renewables could keep growing. Some nuclear projects could deliver. Carbon capture could reduce emissions from gas plants. But we should not assume success by default. Energy systems have inertia. Permitting fights are real. Costs matter. Politics shifts. Many "solutions" depend on policy choices: tax credits, regulatory standards, enforcement that survives changes in administration.

Some climate advocates reject this entire framework. Their argument is simple: we have a carbon budget. Every ton matters. The question is not whether gas is cleaner than coal, or whether nuclear might arrive later. The question is whether any justification—even AI—is worth burning the budget. This position has uncomfortable implications: it would mean slowing the buildout, rationing compute, or accepting that some AI applications are not worth their carbon cost. That is not how the industry thinks. But it is how some climate scientists think. And they are not obviously wrong.

6.8 The Lock-In

A gas turbine installed to serve a data center in 2027 does not shut down when a solar farm or a reactor comes online later. Gas plants operate for thirty years or more. Infrastructure built as a stopgap can still be running in the 2050s, locking in emissions long after the original justification has expired. Every turbine commissioned today makes long-term carbon commitments harder to keep.

The likely future is not a clean transition. It is a messy portfolio: gas running alongside growing renewables, nuclear arriving unevenly, carbon accounting filling the gaps that physics leaves open.

Elena knows this arithmetic. In the control room, the reactor hums. She is proud of the work it took to bring it back. She also knows that a

CHAPTER 6. THE GENERATION

gas turbine down the road was generating power years before the first neutron split. The mismatch is not a failure of planning. It is a fact about how energy systems change: slowly, unevenly, and never as cleanly as the announcements promise.

The student in Ann Arbor has long since received an answer. Coffee has gone cold. The conversation moves on to follow-up questions and clarifications, the easy back-and-forth that feels like talking to a patient tutor. What the student does not see is the physical adjustment that happened somewhere in the regional grid: a gas turbine ramping, fuel flowing through pipelines, carbon dispersing into the atmosphere.

We have traced the token through the stack: mathematics, silicon, racks, buildings, transmission lines, and now generation. Electricity is real. Carbon is real. Water evaporating from cooling towers is real. Every query has a physical cost distributed across infrastructure that spans continents.

But this infrastructure needs land. The Saline campus sits on hundreds of acres that grew corn and soybeans a generation ago. How does farmland become the destination for a question asked in a coffee shop? That answer is not in physics. It is in land. In the next chapter, we go there.

CHAPTER SEVEN

The Land

A realtor's sign still stands at the edge of the driveway. It no longer matters. Three generations of *Harold*'s family farmed this land in Saline Township: corn and soybeans on five hundred acres, soil names he knew by feel before any county survey existed. Now a data center developer owns it all. *Harold* signed before the township ever voted.

Harold is seventy-three. His hands show it: thick knuckles, cracked skin, an old scar across his palm from a combine auger. His grandfather bought this farm in the 1920s. His father expanded through the Depression, buying foreclosed neighbors' land at prices that haunted him for decades. *Harold* survived the farm crisis of the 1980s. He remembers auction signs. He remembers the silence where tractors used to run.

He raised two children here, but both moved away. Chicago. Seattle. Neither wanted the early mornings, the breakdowns, the weather anxiety, the margins so thin that one bad year could wipe out three good ones. He understood. He wanted them to have easier lives.

"They weren't going to farm it," *Harold* says, standing at the edge of his former property. The wind smells like construction now, not soil. "And the offers kept getting bigger. First one price, then double, then triple. I held out as long as made sense." He pauses and looks down at the ground his grandfather cleared. "Maybe longer."

CHAPTER 7. THE LAND

What *Harold* does not say, at first, is the part that makes this sale feel different. The night he signed, he could not sleep. He walked the fields at 3 a.m., frost crunching under his boots, trying to memorize every contour. The rise where his father taught him to drive a tractor. The low spot that flooded every spring and grew the sweetest corn. The oak at the property line where he carved his initials as a boy, then his children's initials, and then nothing—because there was no one left to carve. The money is real. Enough to pay off loans and set up grandchildren he rarely sees. Still, he wakes some nights thinking he hears the combine running.

A local intermediary delivers the first offer, the way land deals often work. Polite phone call, a meeting at a kitchen table, a buyer who does not say much about what he is buying for. Then the offers rise. Not by a little—by enough that *Harold* has to stop and ask himself whether he is being stubborn for the sake of pride. He brings in a lawyer. He reads clauses about options, timelines, confidentiality. He hears words that do not belong to farming: due diligence, escalation, assignment. The developer wants the right to buy later, after more parcels are assembled and power agreements look plausible.

Harold understands the logic. He has assembled acres himself, one neighbor at a time. But there is a difference between buying a field and buying a future. When he finally signs, he tells himself a story farmers have told for generations: the land does not love you back. You work it, and then you hand it on, and the world changes around it. Still, this handoff feels like a severing.

7.1 The Farmland Paradox

Not every farm sells. A few miles away, a centennial farm still operates. A metal plaque by the road marks a hundred years of family ownership. The town has already changed around them: subdivisions, strip malls, the slow retreat of agriculture. Now the next change arrives. Data centers.

What puzzles people is not that land is being sold. Land has always been sold. The real question is *which* land sells. Abandoned factories and idle industrial sites sit a few miles away—the kinds of places politicians and planners say should be reused. Those sites sit empty while clean farmland gets paved. That contradiction has a name in the industry now: the farmland paradox.

The paradox has a rhythm. First comes quiet land buying: options, not deeds, one parcel at a time, paperwork filed under new LLC names—Limited Liability Companies that hide the real buyer's identity—that do not ring bells. Developers do this for a reason. If everyone knows what is coming, prices jump. Neighbors start calling each other. Opposition organizes early. The schedule slips. So buying happens quietly until enough land is assembled to make the project plausible. Then the announcement arrives all at once: a press conference, a rendering, a number so large it does not feel real.

For the people who live there, the story can feel backwards. The decision seems made before anyone speaks. That is part of why land deals produce mistrust, even when the checks clear.

7.2 Why Brownfields Fail

America has thousands of brownfields: steel mills, coal plants, refineries, factories. Sites that carry the scars of the last industrial era, many in communities that lost jobs decades ago and would welcome investment. Policy says we should reuse them. Environmental advocates agree. Local communities say they want their abandoned sites brought back to life. Yet again and again, the market does the opposite. It chooses cornfields.

Understanding why requires accepting an uncomfortable truth: siting is political even when it pretends to be technical. The political theorist Langdon Winner argued in 1980 that "artifacts have politics"—that the technologies we build and the infrastructure we site are never neutral. They encode power relationships, distribute costs and benefits, and those distributions can last longer than the people who made the decisions. Building on greenfield land rather than a brownfield site is not only a real estate choice. It concentrates costs on rural communities and directs benefits toward corporate shareholders and distant users.

Developers have a simpler explanation. Farmland is cheap, clean, fast, simple. Brownfields are expensive, contaminated, slow, complicated. No policy preference has overcome the advantage of speed and certainty. Power infrastructure starts the story, as we saw in the grid chapters. Land is where the story turns into conflict.

The economics are counterintuitive. Midwestern farmland might sell for five to ten thousand dollars an acre. Industrial brownfield can cost ten times that. But at the scale of a multi-billion-dollar data center, even expensive land is a rounding error. *Harold* received what felt like a fortune for his five hundred acres. For the developer, the entire land purchase barely registered against a project budget in the billions. Developers do not choose farmland because it is cheap. They choose it because it is fast.

The real cost of a brownfield site is not the price per acre. It is the months or years lost to contamination studies, title disputes, regulatory approvals, and infrastructure rebuilds. In an industry racing to deploy chips before they become obsolete, time is the binding constraint. Land is not just a surface. It is a schedule.

Why are brownfields slower? Four structural penalties explain the gap.

First: contamination. Brownfields are brownfields because something happened there. Heavy industry leaves behind chemicals, heavy metals, asbestos, coal ash, solvents. Cleaning it up takes time and money. The legal risk can be brutal. Superfund—the federal law requiring anyone who dumped hazardous waste at a site, or currently owns it, to pay for cleanup—can make present and prior owners responsible. Liability stretches across decades. Even when a developer is willing to remediate, they face a hard question: if we buy this site, do we inherit a legal nightmare we cannot price? A cornfield does not have that problem.

Contamination is not only a technical problem. It is a legal and financial one. Disturb contaminated soil, and you can inherit obligations that lenders and insurers refuse to touch. Remediation plans change midstream. Costs jump. Timelines stretch. A site can become a lawsuit factory. Developers can manage these risks, and some do. But when the market is racing, the safer move is to avoid the risk entirely. Buy clean land instead.

Second: infrastructure fit. People assume old industrial sites must have "power." Sometimes they do. But the scale is wrong. A twentieth-century factory might have used five or twenty megawatts. A hyperscale data center wants five hundred megawatts, a gigawatt, more. Existing infrastructure is not "almost right." It is wrong.

Getting new power to any site means applying to the regional grid operator and waiting in an interconnection queue—a formal line of projects requesting permission to connect to the transmission system. As we saw in the grid chapters, that line is years long. A brownfield site's old factory connection does not let a developer skip the wait. The original interconnection agreement expired when the factory closed. A data center developer requesting hundreds of megawatts joins the back of the line, same as on a greenfield. Worse, legacy infrastructure can slow the rebuild. Old equipment must be decommissioned. Underground utilities—gas, water, sewer—must be located and moved. Existing easements and right-of-way agreements must be unwound. The supposed power advantage often turns into a penalty.

Water follows the same pattern. A factory site might have water service, but not at the volumes some cooling systems demand. A gigawatt facility can require city-scale water delivery if it relies on evaporative cooling. That means new mains, new easements, new fights.

Third: complexity. Farmland often has simple ownership—one family, maybe two. Deed histories are clean. Title clears fast. Brownfields have layers: bankruptcies, corporate successors, tax liens, environmental liens, easements, pension claims. Clearing title can take months or years. Until title is clear, you cannot build.

Politics compounds the problem. Communities with abandoned industrial sites want comprehensive redevelopment: housing, retail, diverse employment. A data center brings tax revenue and a limited number of permanent jobs. It rarely feels like a revival. That expectation gap creates friction. Public hearings grow contentious. Officials add requirements: community benefit agreements, local hiring mandates. From the community's view, these demands make sense. From the developer's view, they add months.

Fourth: security. Hyperscalers treat their AI facilities as high-security installations. They want a defensible perimeter, clear sight lines, controlled access points, and space for patrols, with no adjacent buildings offering a view into the yard, no shared walls, no odd geometry. Farmland offers all of that naturally. An urban infill site often does not. The preference is technical, but it has a political effect: it pushes infrastructure away from dense areas and into rural ones.

Put these four penalties together and the timeline gap is consistent. Greenfield sites break ground in months. Brownfields require years. In an industry racing to deploy chips before they become obsolete, years are fatal.

7.3 Policy Fails

If farmland advantages are so strong, why has policy not flipped them? Most brownfield incentives are too small to matter at hyperscale. Federal and state programs exist, but they nibble at the edges. The federal tax deduction that once helped brownfield remediation expired more than a decade ago. EPA grants fund assessments, but a grant that feels meaningful to a small developer becomes noise in a project with billion-dollar capex. Energy-focused bonuses often apply to power generation, not to the data center itself.

Meanwhile, state competition runs the other direction. States offer data center tax exemptions because they are politically easy: a promise of investment and revenue with the cost hidden in tax breaks and higher utility bills for everyone else. Those incentives speed farmland conversion without changing the barriers that make brownfields slow. Policy subsidizes the *result*—more buildout—without paying the price required to change the *pattern*. Developers are not choosing farmland because they enjoy fighting township boards. They choose farmland because it is the only way to hit the schedule.

The barriers are real, but they are not laws of nature. Contamination can be remediated with public funding. Interconnection queues can be expedited for sites with existing grid access. Title complexity can be streamlined through dedicated brownfield review processes. Other countries have done versions of this. The United States has not tried at the scale the moment demands. That is a policy choice, not an engineering constraint.

7.4 STATE BY STATE

The farmland paradox plays out across the country in different ways. Kansas shows the greenfield model in its purest form: flat land, clean title, fast permitting, a utility willing to commit to an aggressive timeline. Farmers receive what feels like a windfall. Land remains a small fraction of the budget. Projects move quickly because nothing about the site slows them down.

Pennsylvania provides the most interesting test of brownfield viability. Some projects try to redevelop old industrial sites. The ones that work tend to have cleaner histories: light industrial, research campuses, printing plants. Not steel mills with slag piles and decades of contamination.

CHAPTER 7. THE LAND

An odd twist: the best grid connections can come with the worst environmental legacy. Old power plant sites sit on transmission infrastructure built for generation, which can be attractive for data centers—even if they require serious remediation.

Arizona shows a different limit: even greenfield development can fail when water becomes the constraint. Some desert sites have the land, the sun, the state-level political support—and still face local resistance. Projects get canceled because residents do not accept the water trade. Hyperscalers walk away, even with big budgets, because community opposition slows the timeline enough to kill the deal.

Michigan became a leading edge of organized resistance in 2025. Projects that might have sailed through zoning five years earlier run into seven-hour meetings, packed rooms, petitions, referendums, moratoriums. Some projects are commercial. Some are backed by public institutions. Opposition grows anyway. The pattern is no longer "find land, file paperwork, build." Siting becomes a political contest.

7.5 Saline's Story

The consent agreement is signed. The lawsuits are settled. *Frank*, who has served on the Saline Township board for eight years, finds his phone keeps buzzing. Voicemails arrive early in the morning. "You sold us out." "Three generations of farmers, and you let them pave it over for computers." *Frank* listens in his kitchen while his coffee goes cold. His wife pretends not to hear. Forty years of marriage teaches you when to give someone space.

Frank understands the anger better than the callers know. His own father farmed until the economics broke him. *Frank* grew up watching farms give way to subdivisions, then strip malls, then the slow decay that comes when young people leave and old people die. His daughter stays closer. She brings the grandkids for holidays. She calls on Sundays. She worries about him the way he used to worry about his father. After the settlement, *Harold* called. *Frank*'s wife said that he sounded lost. *Frank* did not call back. What would he say? That the money was good? That the decision was rational? That the township could not have won? That sometimes you lose even when you do everything right? None of those sentences feel like comfort.

Frank meets *Ellen* for coffee sometimes, at a diner where farmers have sat in the same booths for decades. They talk about nothing and everything. She never asks him about the vote, or whether the consent agreement was the best they could do. He is grateful for that.

Ellen is his neighbor—the farm adjacent to the project site has been in her family for three generations. Through her kitchen window, where she used to watch deer cross the back field at dawn, she now sees excavators and concrete forms. She was twenty-six in 1985, the year the farm crisis hit bottom. She remembers the auction signs. The Hendersons, three miles down the road, borrowed against their land when corn was three dollars a bushel and lost everything when it fell below two. Bankers had been happy to lend when prices were rising. Then prices fell and those same bankers took the land. Mr. Henderson drove his truck into the pond behind the barn. Nobody called it suicide. Everyone knew.

That was the thing about money people. They made their cut on the way up and they made their cut on the way down, and the people left holding the land or the debt or the worthless shares were always somebody else. It had been true in 1985, true in 2008. It would be true the next time, whenever the next time came.

Why did Saline attract the project? Every factor we have discussed lines up. Transmission corridors cross the township. The utility is willing to supply massive capacity. The land is clean, flat, and available. Title clears quickly. No contamination hides under the surface. *Harold* and his neighbors sell at prices that seem generous by farming standards and trivial by data center standards.

From the developer's view, Saline is simple. From the community's view, nothing is simple. Once land is assembled and power access looks plausible, bargaining power shifts. The township can say no, but saying no can trigger a lawsuit the township cannot afford. So the consent agreement becomes the compromise: protections for wells, noise limits, money for a community fund, investments in the fire department. Those protections are concrete. The existence of a consent agreement is itself a sign of a new reality: communities no longer accept vague promises. They demand specific terms.

7.6 WATER

Land is one constraint. Water is another. Cooling at AI scale, as we saw earlier, involves trade-offs among energy, capital cost, and water. Evaporative systems are efficient but consumptive: water turns into vapor and leaves the system. Closed-loop systems use less water but add cost. At gigawatt scale, the volumes become city-like.

Industry advocates frame this as a non-problem. Nationally, data center water use is modest compared to agriculture or even golf courses. In aggregate, that is true. But the comparison is misleading. Golf courses are dispersed across thousands of locations. Data centers cluster. A single large facility can consume millions of gallons per day, and that demand peaks in summer, exactly when drought risk is highest. What looks manageable as a national average can overwhelm a single county.

In Saline, the project describes its water use as closer to a large office building than a thirsty factory. That claim may hold for one facility. But communities remember drought summers and wells going bad. They know that if a region becomes a data center cluster, one facility turns into five. The fear is not about a single project's numbers. It is about what comes next.

7.6. WATER

In Northern Virginia, water demand rose sharply as data centers multiplied. In the Southwest, aquifers were already strained before the first data center proposal arrived. In Arizona, farmers like *Roberto*—whom we met watching his wells drop outside Phoenix—see irrigation allocations shrink while data center permits get approved. The water that grew crops becomes the water that cools servers. They do not experience a national average. They experience a competing claim on a resource that is already shrinking.

There is also a change that has nothing to do with cooling. Farmland and forest absorb rain. Concrete does not. When hundreds of acres of soil become impervious surface, groundwater recharge drops and runoff increases. Drainage patterns that served the land for centuries change in a single construction season. The cooling towers are the visible water story. The lost soil is the invisible one.

Community impact is not only environmental. Data centers bring tax revenue, construction jobs, a few hundred permanent positions in operations and facilities work. But they also bring noise from cooling equipment and backup generators, truck traffic, a change in the character of land. And they bring a new kind of politics: debates over who benefits, who pays, who decides. Whether those benefits outlast the disruption is a question for later—or never.

Environmental justice advocates see a familiar pattern. The burdens of industrial development—noise, traffic, water strain, visual disruption—land on communities with less political power. Benefits diffuse: tax revenue spreads across a state, shareholders collect returns globally, users type queries from distant cities. The community hosting the facility experiences the transformation directly. Everyone else experiences it as abstraction. This is not unique to data centers. It is the recurring geography of American infrastructure: highways through Black neighborhoods, refineries in poor parishes, waste facilities in counties that cannot afford lawyers. Data centers follow the path of least resistance. That path often leads to places already stretched thin.

Some communities want the revenue. Others value land and quiet more. Some have the legal and financial capacity to fight. Others do not. No single answer exists. But a clear trend emerges: opposition grows. Organizing gets better. Projects face ballot initiatives, moratoriums, referendums, lawsuits. Siting is no longer a paperwork step. It is a political contest.

7.7 What Comes Next

So what happens next? The farmland paradox will persist. Its drivers are structural, not accidental. Greenfield sites are faster, cleaner, simpler. But three trends can reshape siting over the next decade.

First, community resistance becomes a real constraint. Some projects will fail because no community wants to host them. Others will succeed by choosing locations where opposition is weaker, by offering concrete benefits, by designing facilities that reduce noise, water use, and visible disruption.

Second, water constraints redirect investment. Water-stressed regions become less viable as cooling demands rise. Water-rich regions look more attractive. That includes the Great Lakes states, the Pacific Northwest, and parts of the Northeast.

Third, brownfield development is niche today, but the barriers keeping it niche are not permanent. Some brownfields already work: sites with clean histories, light industrial pasts, or former power plants sitting on transmission infrastructure built for generation. The barriers that block the rest—contamination risk, infrastructure gaps, title complexity—are each addressable with the right policy tools. Tax preferences, expedited permitting, remediation funding, and dedicated review processes could close the gap between a contaminated industrial site and a clean cornfield. That gap will not close on its own. Whether policy rises to meet it is a question we return to in the final chapters.

A fourth trend: learning. Townships read about each other now. A board member in Michigan can watch a Virginia hearing on YouTube. An activist in Arizona can share a noise ordinance template with a county in Georgia. Communities borrow tactics and language. Developers borrow counter-tactics. Everyone gets faster. Siting becomes harder to treat as a one-off local dispute. It becomes a repeating national pattern.

By early 2026, construction is underway. *Harold* drives past his former land sometimes. His daughter wishes he would not. He takes the long way to the grocery store, the route that passes the property line where his grandfather planted oaks now scheduled for removal. Survey stakes. Early grading. Signs announcing what is to come. Security fencing where the farmhouse used to stand.

"It's strange," he says, sitting in his pickup truck with the engine idling against the winter cold. "Farming that land for fifty years. Watching my father farm it before me. Watching my grandfather plant those oaks because he wanted shade for generations that hadn't been born yet." He pauses. "And now it'll be something completely different. Something I don't even understand."

His daughter calls from Chicago. How is he doing? Fine, he tells her. He does not mention the dreams where he is still farming, or the way he sometimes catches himself planning next year's rotation before remembering there will be no next year.

"But the world changes," he says, looking out at the construction equipment, the men in hard hats, the transformation underway. "The world always changes. I just wish I knew what."

CHAPTER 7. THE LAND

Land, in the end, is not the most expensive input to the AI buildout. But it is the most permanent. Once farmland becomes a campus, it does not go back. The footprint extends beyond the campus fence. Transmission corridors cut through forests and wetlands. Access roads fragment habitats. Substations occupy land that was never counted in the original site plan. The full environmental cost of a data center includes not just the acres it occupies but the infrastructure that serves it. The question is not only where these facilities can be built, but who decides—and how those decisions get financed.

In the next chapter, we follow the money.

CHAPTER EIGHT

The Money

On September 4, 2024, Blackstone announced it was buying AirTrunk, an Australian data center operator, for sixteen billion dollars. It was the largest data center deal in history.

David watches the announcement from his office in Manhattan, high above Park Avenue. The September light slants through glass, his coffee has gone cold, and the floor is quiet in the specific way finance offices are quiet: thirty people, thirty screens, no one talking, just the soft hum of machines. Fifteen years in private equity, eight focused on data centers, and he has never seen anything like this. Sixteen billion dollars, at a premium high enough to make his colleagues gasp.

David pulls up his models, the familiar grid of assumptions that has guided a decade of decisions. If Blackstone is paying this much for assets on the other side of the world, what does that imply for American facilities? For the deals his fund is circling? The math says one thing. His gut says another—a tightness below his sternum that he has learned to trust. This is either the beginning of the greatest infrastructure investment cycle in history, or the top of a market that will make fools of everyone who chases it.

———

CHAPTER 8. THE MONEY

Blackstone is not just buying a company. It is placing a bet that data centers are no longer a niche corner of real estate but the backbone of the next economic era. Leadership says the quiet part out loud: the world is heading into a trillion-dollar buildout in the United States, and another trillion elsewhere. Blackstone intends to be a central financial investor in that buildout.

The analogy they reach for is familiar: the electrification of America. Technology arrives first, then concrete and copper follow. Infrastructure has to be built by someone. Increasingly, that someone is private capital.

8.1 Trillion-Dollar Scale

Scale this large is hard to hold in the mind. By the end of 2025, documented investment in U.S. data center infrastructure had exceeded a trillion dollars across more than six hundred projects. Average projects now measure in billions, not millions.

David keeps a spreadsheet, the kind of document that grows like a living organism. Rows and tabs multiply. The numbers stop feeling like numbers and start feeling like weather. He updates it on Sunday nights, after his kids are in bed, when the house is quiet enough that he can think. A new campus announced in Texas. A utility filing in Virginia. A land buy in Michigan. A rumor that becomes a permit, a permit that becomes a financing.

When *David* started in this business, a five-hundred-million-dollar deal was a big deal. Now five hundred million feels like a rounding error. "I tell my kids I work in finance," he says. "They ask what I finance. I say data centers. They ask how big. I say a trillion dollars." He shrugs. "They think I'm making it up." He is not. The curve steepens quickly, and that steepening changes who can play.

8.2 How the Market Changed

To understand the money, see the market in phases. First came the REIT era. A REIT—Real Estate Investment Trust—is a company that owns real estate and lets investors buy shares, the way you might invest in an apartment building. Data center REITs like Digital Realty and Equinix treated data centers as specialized real estate: long-term tenants, predictable cash flows, steady rent escalations. These public companies grew into stable giants. A data center was an industrial building with a lot of power. The world this era assumed was slow. Compute grew, but not explosively. Rack densities rose, but not by orders of magnitude. Power procurement was a headache, not the headline.

Then that world ended. Private equity arrived, signaling that the sector was being revalued. Big firms paid large premiums to buy public platforms and run them privately: more aggressive development, less quarterly pressure, more willingness to build ahead of demand. The deals got larger, the prices got higher, and the logic shifted from "steady yield" to "growth at speed."

Then came the AI transition. ChatGPT launched, and suddenly everyone believed the demand curve was about to bend upward. AI was not one new tenant. It was a new class of tenant that wanted power and space at scales that strained the old industry.

Then came the mega-deal era. Project sizes got too big for any single balance sheet. Consortiums became normal. Capital stacks became deep—seven parties, three time zones, contracts thick as phone books. No longer just a real estate deal. An energy deal, a supply chain deal, a political deal, a timing deal where being six months early can matter more than being six percent cheaper.

David describes a recent negotiation. The parties argued for hours about one clause: the power commitment guarantee. If the utility cannot deliver on time, who bears the risk? The operator? The tenant? The investors? The lenders? He laughs, but it does not sound like amusement.

"We settled at two in the morning," he says. "I'm not even sure what we agreed to. The lawyers are still drafting."

8.3 The Capital Stack

So who provides the capital? Start with the hyperscalers: Microsoft, Amazon, Google, Meta. Staggering sums on data center infrastructure, not as passive investment but because compute is strategy. Market position is the return they care about. Their spending also enables everyone else—when a hyperscaler signs a long-term contract, that contract becomes the foundation lenders can underwrite. A speculative build becomes something that can raise debt at scale.

Then come private equity and infrastructure funds. Data centers look like an asset class to them: long-lived infrastructure with contracted revenue and demand that seems durable. Money is only part of what they bring. Operating playbooks matter too—acquire platforms, improve operations, accelerate development, exit. Scarcity drives the logic. The best power positions are hard to secure. The best interconnection queue spots are already claimed. The best operator talent is booked. In a scarcity market, owning the platform matters.

Public REITs remain important even as some go private, providing liquidity, scale, and market benchmarks. Sovereign wealth enters the picture too. Projects reaching tens of billions need capital with long time horizons and deep pockets. Call it "quiet money": pension funds, endowments, insurance pools. Stable returns, not drama. Not chasing the next app. Buying into the physical layer.

This is the financial map of the AI buildout: hyperscaler demand, private capital structures, and public systems that ultimately bear the externalities.

8.4 How Deals Get Done

Deal structures matter as much as the deals themselves. Each structure solves a different problem. Take-privates: when a public company's stock price does not reflect the value of its land and power positions, private equity buys it, takes it off public markets, builds aggressively without quarterly earnings pressure, and sells later at a higher price. Joint ventures: when a single deal is too large for one fund, two or three sponsors share the risk and capital. Sale-leasebacks: when a developer needs cash for the next project, they build a facility, sell the real estate to an investor, then lease it back—freeing up capital without losing operational control. Vendor financing: when hardware providers like NVIDIA want to lock in demand, they extend payment terms that reduce near-term cash needs for buyers.

But the single most important element in many deals is the anchor tenant—a hyperscaler that signs a long-term contract before construction begins. *David* is blunt: "Without an anchor, you can't get a loan. Period." Banks do not care about growth projections. Counterparty risk is what matters. Who is on the other side of the contract? Microsoft is investment-grade. Some startup no one has heard of is not. So the anchor tenant becomes the bridge between Wall Street and concrete.

Then the rest of the stack gets built around it. One sponsor rarely carries the whole equity check—two or three funds share it. Debt comes in layers: conservative senior loans on top, then more expensive money underneath for the parts lenders will not touch without extra yield. Sometimes hardware vendors help on terms, because it pulls demand forward and locks in a customer relationship. Utilities enter the stack too, in their own way. A power contract can function like a kind of financial backbone, promising delivery, or payment, or both. In the mega-project era, a single campus can require a small army of lawyers just to keep the agreements from colliding.

CHAPTER 8. THE MONEY

David jokes that on the biggest deals, paperwork is part of the plant. The friction matters. Every extra party adds another set of incentives, another set of exit timelines, another way a project can stall—not because the soil is wrong, but because the signatures do not line up. When we hear "a fifty-billion-dollar campus," part of what we are hearing is concrete and copper, and part of what we are hearing is coordination. On some deals, the legal work alone runs into the tens of millions, because the contracts have to allocate risk among parties with different timelines and different fears. Who pays if power is late? Who pays if the tenant downsizes? Who gets control when the project needs more capital? When the answers are unclear, money gets expensive. So the agreements get long. This is what it can take to finance a campus at the scale of a small city.

Here is where valuation gets strange. In the REIT era, valuation looked like real estate: buy contracted cash flows, assume modest growth, earn returns through stable income and gradual appreciation. In the mega-deal era, prices begin to incorporate something else: optionality. Sometimes that shows up as simple sticker shock—multiples that once sounded like real estate start to sound like software. Twelve times earnings becomes twenty, then thirty, then fifty or more.

Investors justify it by saying they are not buying a warehouse. They are buying a platform—an operating company with secured access to electrical power, land entitled for development, and relationships with utilities and tenants. That combination is scarce and hard to replicate. A new competitor cannot simply decide to enter the market; it would need years to assemble the same position. So the price reflects not just today's rent rolls but the ability to sign future tenants at future rates, in a market where power is the bottleneck and supply cannot keep up with demand. The investor is paying for what the platform can become, not only what it is—betting that demand for AI computing will keep climbing and that owning a scarce position today will look cheap in hindsight.

David says it plainly: "Now we're pricing things that don't exist yet. Capacity that hasn't been built. Demand that might not materialize. It's educated guessing dressed up in spreadsheets." On its own terms, the logic holds together. If demand keeps climbing fast and supply stays constrained, today's expensive capacity can look cheap in a couple of years. Pay a premium now, collect rising rents later—the math works if the assumptions are right.

The problem is the assumptions. Demand has to keep growing. Supply has to stay tight. Computing architectures cannot shift in ways that strand today's facilities. Power has to arrive on schedule. Each assumption is plausible on its own. But they all have to hold at once, and no one can verify any of them in advance.

And because prices are high, there is not much cushion. If power arrives late, the building sits idle while interest accrues. If rates stay high, debt costs eat returns that used to look easy. If rents do not rise as fast as everyone expects, today's premium can turn into tomorrow's regret. This is why the mega-deal era feels so tense inside finance offices. The upside is enormous, but when the price already assumes the upside, the room for disappointment is small.

Private markets can hide this longer than public markets. Public stocks reprice in days; private valuations can drift for quarters, because appraisals lag and deals are infrequent. But the math does not care about the reporting calendar. Are these valuations justified, or are we watching bubble pricing? *David*'s honest answer is the only honest answer: "Nobody knows."

———

Saline illustrates these dynamics in practice. Related Companies, the developer, is a huge real estate firm famous for projects like Hudson Yards in Manhattan. Its founder, Stephen Ross, is a billionaire with deep ties to Michigan. Related Digital is the company's data center division. Saline is its biggest bet yet. The logic is familiar: skills from mega real estate projects should translate into data center development—land assembly, construction management, tenant relations, capital formation. Should.

David's fund looks at Related Digital and passes. "Too early," he says. "Related knows real estate. They don't know data centers. Not yet." The judgment is not moral. It is financial. Investors price execution risk. This is part of why brand names matter—in a market where so much depends on timelines and competence, a developer's reputation can be worth real money.

Saline is also a consortium story. Oracle serves as anchor tenant. OpenAI's involvement signals high-value AI workloads. Utility power commitments are essential. The state provides incentives. The township bears impacts. Finance does not sit above the physical world. It plugs into it.

8.5 The Downside

Then comes the question that keeps *David* awake: What could go wrong?

Start with demand risk. The entire thesis assumes AI computing demand keeps growing fast. But efficiency gains could outpace that growth, flattening the curve. Competition might compress AI service pricing. Economic slowdowns could delay enterprise adoption. Any of these could leave facilities built for 2028 sitting partly empty until 2032. Or forever.

Technology poses another threat. Today's facilities are optimized for GPU-based computing, but computing history is a history of architectural transitions. When the dominant architecture shifts, specialized AI facilities could lose value faster than investors expect. General-purpose data centers serve many workloads. AI-specific builds have fewer alternative uses. "We're building for a technology that's six years old," *David* says. "What if something better comes along?"

Policy risk looms large. Export controls shift. Energy policy shifts. Trade policy shifts. Rules change between administrations, sometimes overnight. Hard to model a world where a regulation can flip and reprice an asset before the next earnings call.

Power risk may be the most concrete. Many announced projects depend on infrastructure that does not yet exist. Transmission lines need rights-of-way. Substations need transformers. Generation needs permits. Delays cascade. Without power, a data center is an expensive building. Not an investment.

Valuation risk compounds everything else. High prices leave little room for disappointment. Sentiment shifts, multiples compress. An asset bought at a high multiple can be worth far less later—even if it is operating well. Public markets reprice in days. Private markets take longer. But the math catches up.

Interest rate risk rounds out the list. Many deals rely on debt. Some debt floats with rates. Rates rise, payments rise, returns shrink. High rates can also freeze new deals entirely—buyers want lower prices, sellers resist, transactions slow to a crawl.

David lies awake thinking about all of these at once.

8.6 PUBLIC EXPOSURE

Most of these risks fall on investors. But data center development also creates exposure for people who did not choose to participate.

Utilities offer one example. When a utility commits to serve a gigawatt customer, it builds infrastructure to deliver that power—transmission lines, substations, sometimes new generation. Those investments can reach hundreds of millions of dollars. If the customer later cancels or scales back, the utility is left with a stranded asset: infrastructure that costs real money to build but has no one to serve.

This is already happening. Utilities have begun culling speculative projects from their pipelines, dropping requests that look unlikely to ma-

CHAPTER 8. THE MONEY

terialize. Some built capacity for customers that never followed through and are now trying to sell unused power back into wholesale markets. In one case, a utility ended up holding roughly seven hundred and fifty megawatts of capacity it had built for a data center customer that never showed up—enough power for a mid-sized city, sitting in the ground, unused.

Utilities respond by writing stricter tariffs: longer minimum contract terms, high minimum payments, exit fees. The goal is to shift risk back to the data center operator. But protection works only if the operator can pay—and does. In some places, the terms are blunt: you want the grid to build for you, sign a decade-plus contract. Use less power than planned, you still pay most of what you promised, because the grid still has to carry the capacity. Think of it as reserving a seat on an airplane. The plane takes off whether you show up or not. These rules are not only technical. They are political. They decide whether a boom is underwritten by a few companies or by millions of ratepayers.

Then there are ratepayers. Grid upgrades built for data centers flow into rate bases that include all customers. Capacity market costs rise. Bills rise. No line item says "AI." Even if a specific data center fails, the infrastructure built to serve the boom still shows up in everyone's rates. In parts of the country, those costs are already visible, even if most people cannot name the cause. The bills just creep upward. In PJM, the independent market monitor points to data centers as a major driver of recent capacity price increases. Economists call rising prices a "market signal," a sign that new generation is needed. But for a family in Ohio or a small business in New Jersey, the signal arrives as a higher electricity bill. The price mechanism works as designed. The people paying the price did not design it.

Municipalities face a different kind of exposure. When a big project is announced, local governments start spending in anticipation. They widen roads, extend water and sewer lines, hire inspectors, expand emergency services. Sometimes they borrow money to pay for these improvements, betting that the project will generate enough new tax revenue to cover the debt. The borrowing takes different forms—bonds, special tax districts, or other arrangements—but the logic is always the same: spend now, collect later. If the project arrives on schedule and pays the taxes everyone expects, the bet works. If the project is delayed, downsized, or cancelled, the debt remains. The roads are already widened. The bonds still need to be repaid. The city has to find that money somewhere else, which usually means cutting services or raising taxes on the residents who were already there. A cautious finance director hears every big announcement the same way: What happens if this is late?

Tax exemptions create a quieter version of the same problem. Many states waive sales taxes on data center equipment to attract projects—exemptions that can be worth tens or hundreds of millions of dollars. That is revenue the state chose not to collect. If the project delivers the jobs and investment it promised, the trade-off may be worthwhile. If it does not, the lost revenue is gone permanently. The state spent money it will never get back on a bet that did not pay off. In each case, the pattern is the same: private investors capture the returns, and public systems absorb the risk.

If this sounds familiar, it should. The telecom boom of the late 1990s offers a cautionary template. Telecom companies invested hundreds of billions in fiber infrastructure, much of it financed by debt. Demand did not arrive on schedule. Fiber sat dark. Companies failed. Workers lost jobs. Communities built around telecom hubs were left with stranded expectations.

The AI data center buildout differs in important ways. Anchor tenants exist. Power is a real physical constraint. Current demand is measurable. But the lesson remains: when private investors bet aggressively on new

CHAPTER 8. THE MONEY

technology, others often share the costs of failure. Not just those who placed the bets.

Money also changes geography. Traditional hubs remain important, but the mega-project era pushes investment toward places with gigawatt-scale power and abundant land. Money follows transmission corridors, fuel supply, cooperative utilities. Kansas. Pennsylvania. New Mexico. Texas. Michigan begins to appear on the map.

The money follows the constraint.

The exposure runs deeper than most people realize. The biggest AI companies—Microsoft, Google, Amazon, Meta, NVIDIA—are not just customers of the data center buildout. They are among the largest companies in the world by market value, and together they make up a significant share of the S&P 500 and similar stock indices. Most Americans with a 401(k), a pension, or any other retirement plan own shares of these companies whether they know it or not, because index funds buy them automatically.

For those who do hold such accounts, this creates a strange circularity. The same buildout that is transforming a rural township is also propping up the retirement savings of some of the people who live there. If the AI boom continues, their accounts grow. If it falters, their accounts shrink. The data center down the road may sit half-empty. The tax revenue the township was counting on may not arrive. The utility may be stuck with infrastructure no one needs. The boom and the bust both land on the same people.

But not everyone has a retirement account or any stake in the stock market. Many Americans—and farmers in particular—have little or no equity savings. Farming is self-employment; wealth is tied up in land, equipment, and operating loans, not index funds. For these families, the calculus is simpler and harsher. They bear the costs of the buildout through

8.6. PUBLIC EXPOSURE

higher electricity bills, local tax shifts, and the transformation of their landscape, but they share in none of the financial upside. If the boom succeeds, their neighbors with 401(k)s see gains. They see a changed community. If the boom fails, everyone loses together.

David's office sits thirty floors above Park Avenue. He looks out at a city made visible by a hundred years of infrastructure investment: subways and bridges, tunnels and towers. Someone paid for all of it. Someone profited. Someone lost.

"You want to know if I think this is a bubble?" He does not turn from the window. "The honest answer is I don't know. Nobody knows." He has seen cycles. The financial crisis. The cloud boom. The crypto bust. Each time, smart people got it wrong. "The difference this time is scale," he says. "The bets are bigger. The consequences are bigger. If we're right, we built the infrastructure for the next fifty years. If we're wrong..." He trails off.

Back in Saline, *Frank*—the township board member who has watched this project from the first land inquiries—sits in a meeting room with beige walls and folding tables. Outside, through windows that need cleaning, he can see the construction site: cranes against a gray sky, earthmovers crawling, a preserved red barn that will soon be the quaintest thing for miles.

Frank thinks about the investors he will never meet, in places like Manhattan, Singapore, and Abu Dhabi, who decided that his township was a good bet. He wonders what they see when they look at their spreadsheets. Whether they picture the farms that used to be here. Probably not. Their money flowed from offices on the other side of the world and became concrete and steel, transformers and generators, cooling systems and fiber. It became the physical substrate of the digital economy. It became his township's future.

Frank did not choose this. Neither did his neighbors. Decisions made in rooms they will never enter, by people they will never meet, based on calculations they would not understand even if someone explained them. But they will live with the consequences for decades. His granddaughter will be middle-aged when the first leases expire. Her entire adult life, spent in server country.

8.7 Who Bears the Risk

The trillion-dollar buildout is not an abstraction. *Harold*'s land deal. *Ellen*'s retirement account rising on the same companies paving over her neighbors' fields. *Frank*'s sleepless nights and *David*'s sleepless nights— different in every particular, alike in their uncertainty. The facilities built today will stand for decades. Either essential infrastructure for an AI-powered economy, or monuments to a moment when investors believed something that turned out not to be true.

No one can say which. What we can say is simpler: money is real, land is real, transformation is real. So the question becomes: who gets invited into the upside, and who gets handed the risk? That is what incentives decide.

In the next chapter, we follow the deals governments make to attract this buildout—and what those deals cost.

CHAPTER NINE

The Incentives

The press conference looks like every other economic development announcement in Michigan. Governor Gretchen Whitmer stands at a podium flanked by executives in dark suits, their company logos arranged on a backdrop behind them. Cameras click in staccato bursts. Reporters scribble notes. But the numbers being announced are not the usual ones.

On December 30, 2024, Whitmer signs Senate Bill 237 into law, creating Michigan's first targeted data center tax incentive program. It is not a grant or a check written from the state treasury to a corporation. It is something quieter: an exemption. If you build a qualifying data center, Michigan says, you do not pay sales and use tax on the equipment that makes the building a data center at all—servers, cooling systems, backup generators, networking gear. On a project that spends billions on equipment, that exemption can be worth hundreds of millions of dollars.

Exemptions spread easily for exactly this reason. They do not look like spending, even though they function like spending. No check gets written on camera. The cost shows up later as foregone revenue, buried in budget documents most citizens never read. No line item says "we paid for the data center." The politics of "not collecting" are far easier than the politics of "paying."

The ink is barely dry before the announcements begin. Michigan goes from being off the list to being on the list. Developers show up. Land agents show up. Power studies pile up on utility desks. Suddenly, Saline Township is not just a local zoning fight—it is part of a statewide sales pitch. The incentive system at work: the political plumbing of the build-out, the hidden terms that turn projects from *maybe* into *approved*.

9.1 Virginia's Template

Why do states do this? The answer starts with a place that became the center of the internet almost by accident. In the early 1990s, government contractors clustered near the Pentagon and CIA in Northern Virginia. They built fiber networks for classified communications and connected to an early internet exchange point. When the commercial internet took off, the infrastructure was already in the ground.

Companies needed space for servers, and they needed it close to the exchange points where packets switched hands. Speed mattered. Once a few data centers appeared, more followed, each new facility making the location more attractive for the next. Interconnection got easier. Vendors clustered. Talent accumulated. Northern Virginia became "Data Center Alley."

Then something else happened. Local officials realized these buildings were full of expensive equipment, and if they charged sales tax on that equipment, developers could pick up and go somewhere else. Virginia offered an exemption. At first, it looked like a local trick tied to a local advantage. Other states saw the template. Once it existed, it spread.

Virginia's version became enormous. By fiscal year 2023, the state was waiving almost a billion dollars in sales tax revenue in a single year. Supporters argue this is the price of being the place the internet lives. Critics argue the lost revenue is the whole point—a subsidy hiding inside a tax code instead of appearing as a line item. The sheer scale sends a message to every other state: if you do not offer something comparable, you may not even get a meeting.

9.2 The Template Spreads

Incentive packages rhyme because they are built from a small set of parts. Sales tax exemptions on equipment are the big one. Property tax treatment is another—sometimes abatements, sometimes special assessment rules, sometimes agreements that phase taxes in over time. Infrastructure support often comes next: a state or locality helps with roads, water extensions, or site work, while a utility commits to new substations on an accelerated timeline.

Energy pricing can be part of it too. Large customers negotiate special contracts because they want predictable rates, and utilities want predictable revenue. Regulators bless the arrangement. Then come the softer incentives: job training programs, workforce grants, expedited permits, a single point of contact at an economic development agency who can make the process feel smooth.

None of this changes physics. Incentives do not create a transmission line, manufacture a transformer, or conjure water out of a drought. But they change who pays, and when. They move money across a spreadsheet in a way that can make a project look inevitable.

Each state pitches itself differently, even when the incentive packages look similar. Virginia sells density—it already has the fiber, the interconnection, the deep bench of workers who have run these facilities for decades. Texas sells power and speed: developers can find huge sites, build fast, and pair with generation in ways that feel almost like building a private utility. Kansas sells urgency, a state with no hyperscale presence that can suddenly assemble a project pipeline by offering aggressive terms and fast timelines.

Pennsylvania sells a different story: reuse what already exists. Old industrial sites, old grid infrastructure, a chance to trade brownfields

CHAPTER 9. THE INCENTIVES

for new investment. In practice, those sites come with problems—contamination, messy titles, slow remediation. Speed usually beats the virtue of reuse in this market. Even the "responsible" pitch can lose, not because anyone disagrees with it, but because the calendar is the deciding factor.

Incentives keep escalating because the logic demands it. One state offers an exemption. A competitor matches it. The next competitor adds something else. Job requirements fall, since data centers create few permanent jobs. Investment thresholds rise, since developers can always promise bigger. Exemption windows lengthen—the equipment spend is massive. At a certain point, the packages start to look the same, and we have to ask: if every state offers the same deal, does any of it matter?

Researchers argue about the answer. Some studies suggest incentives can sway location choices at the margin. Others suggest they mainly affect timing—a company that was going to build anyway builds sooner because the deal is too good to pass up. Still others suggest incentives are mostly a transfer: money that would have been tax revenue becomes private return. Regardless of the research, states behave predictably. No one wants to be the first to stop. Stop while your neighbor keeps offering deals, and you worry the investment will cross the border. Everyone keeps playing, even when they admit, in private, that it feels like a race nobody can win.

Economists have a name for this structure: a prisoner's dilemma—a situation where everyone would be better off cooperating, but each individual does better by defecting, so everyone defects and everyone loses. Every state would be better off spending less on incentives, but any single state that unilaterally stops can lose investment to states that keep offering deals. Changing the pattern would require coordination. That is hard in American politics, where governors compete and legislators guard local interests. No one wants to be accused of "losing jobs" because they refuse a subsidy another state is happy to provide. The incentives keep flowing, and the real negotiation shifts to the details: who qualifies, what gets disclosed, who pays for grid upgrades, and what happens if promises are not met.

David has seen this from the other side of the table. His fund has lobbied for incentives in Virginia, Georgia, and Texas. They pay consultants who know which officials to call. They join industry groups that argue for favorable treatment. The polite work of persuasion.

"I am not proud of all of it," he admits. "We are very good at getting states to compete against each other. That is literally our job." No swagger in his voice. He says it like someone describing gravity. His investors want returns. Incentives improve returns. Clean logic.

That clean logic creates an ugly question: if the system is designed for investors to squeeze the best deal out of governments, is it designed for communities to do anything but react? *David* has watched the debates over ratepayer protection, the fights about who pays for grid upgrades. "The system works for us," he says. "I am not sure it works for everyone else."

9.3 Michigan Joins In

Before Senate Bill 237, Michigan was simply not in the game. Site selection consultants did not put the state on short lists. When developers compared packages, Michigan looked expensive on paper. Legislators borrowed the template. They designed SB 237 to look familiar to the industry: a sales and use tax exemption tied to thresholds, a state agency that designates eligible zones, modest job requirements, and high expected wages.

The bill moved quickly. *Frank* watched part of it from his kitchen table in Saline Township. October 2024, and the bill was still working through

CHAPTER 9. THE INCENTIVES

committees in Lansing. On his laptop screen, people in suits talked about "investment" and "competitiveness." Michigan's workforce. The future.

That same day—a year before the board vote, before any of them knew what was coming—*Harold* called him, voice tight. What should he do about the men in suits who kept showing up at his farm? They kept raising their offers. *Harold* had never seen this kind of attention for his land. *Frank* had no clean answer. He could not promise the project was real, or that it would be good, or that it would be stopped. He could only tell his friend what he himself was learning: when a project gets big enough, it starts moving before anyone votes.

Economic development officials testified about jobs and investment. Utility representatives talked about capacity. Industry representatives talked about Michigan's land and workforce and proximity to research universities. Critics raised concerns too—emissions, water use, ratepayer risk, transparency. The bill passed. The signal was sent: Michigan wants data centers.

Michigan carried old scars from incentive programs that went wrong. The state had tried big subsidy experiments before, and the results were mixed—some programs created long obligations that outlasted the political moment that produced them. Even as Michigan adopted the data center template, it tried to build in guardrails: thresholds, eligibility rules, the promise of accountability. The problem is time. A data center deal moves from announcement to construction in months. Measuring whether an incentive program actually delivered on its promises—the jobs, the investment, the tax revenue—takes years, sometimes a decade or more. By the time anyone has enough data to judge whether SB 237 was worth it, the legislators who voted for it have left office, the governor has moved on to the next ribbon cutting, and the facilities are already built. The guardrails may be real, but accountability arrives long after the decisions are irreversible.

Cause and effect get tangled here. SB 237 did not create the Saline Township project from nothing. Related Digital was already buying options on farmland. Power studies were already underway. The machine was already moving. What the incentive did was change the terms of movement: it lowered costs, told the market this project had political tailwinds, and gave the governor a podium moment—a way to claim the buildout as an intentional choice rather than something that just happened.

It also put township officials in a new position. No longer just weighing a development proposal. Now weighing a project the state had already decided to celebrate.

9.4 The Democracy Deficit

Frank understands township zoning. For years, he has read site plans, argued about drainage, asked questions about setbacks and parking. Then a proposal arrived that did not fit his mental categories. Billions of dollars. Hundreds of pages of technical documents. A power demand that made him look up comparisons, because the number did not feel real.

He spent a weekend trying to understand what he was being asked to approve. His degree in mechanical engineering, earned forty years ago before a career in local manufacturing, helped a little. But the scale defeated him—not because he was careless, but because the project was built from a kind of knowledge township boards are not designed to hold. Science and technology scholar Sheila Jasanoff calls this a "democracy deficit": the gap between the decisions citizens must make and the knowledge required to make them well.

That deficit shows up in room after room across the country. Township boards are asked to decide whether a gigawatt-scale facility is good for their community. State legislators vote on incentive programs based on projections they cannot independently check. Everyone must act quickly, because the companies always claim urgency. In that rush, trust becomes a substitute for analysis.

Frank is not alone. Local boards across the country receive documents that assume a working knowledge of terms like interconnection queue position, PUE, redundancy levels, thermal envelopes. These are not bad-faith terms—they are the language of the industry. But when a board member asks, "What does this mean for our water? For our roads? For our electric bills?" the answer often comes back as a mix of reassurance and abstraction. Abstraction is a hard thing to vote on.

9.5 The Jobs Math

Every press conference includes job numbers—that is the language of economic development. But data centers do not behave like factories. A factory employs hundreds or thousands of workers to operate production lines. A data center employs relatively few people to oversee vast amounts of equipment that mostly runs itself. Saline, at full buildout, expects roughly four hundred and fifty permanent jobs on an investment of roughly seven billion dollars. That is about fifteen million dollars per permanent job, compared to a few million for a typical factory. The gap reflects what a data center actually is: a building full of machines tended by a small, highly skilled workforce.

Proponents respond in predictable ways. The jobs that exist pay well. Indirect jobs follow: security, maintenance, food service, contractors. Tax revenues, especially property taxes, can transform school budgets and local services. Proponents also argue that a data center hub attracts other firms—suppliers, contractors, satellite offices—though evidence for this cluster effect in data centers is thin compared to industries like automotive or biotech, where supply chains are local and interdependent. The stronger argument is construction work: a huge campus can employ thousands of union workers during the build. Real paychecks, real apprenticeships, real mortgage payments. Talk to people in the trades and you hear a practical view: we build what gets built. The critique that the permanent headcount is small does not pay this year's bills.

Critics are not wrong either. State budgets are finite. Spend public capacity on one kind of development and you cannot spend it on another. The economics are hard to verify because impact studies are often written by consultants hired by the same people who want the project approved. Governors cite jobs, skeptics cite cost, and the decision is made in a fog.

9.6 Who Pays the Bills

Now the hardest part: electricity bills. When a utility builds new infrastructure for a customer, someone pays. In the old world, big customers were rare, and their needs were modest. A new substation served a factory. The cost spread across a million customers and no one noticed.

In the AI world, a single customer can demand a gigawatt. A new line. A new substation. Upgrades across a region. Huge costs. In the traditional model, a fifty-million-dollar upgrade spreads across millions of customers, and the per-household increase is small enough to disappear in the noise. At gigawatt scale, the numbers stop disappearing. If the grid is upgraded for a data center that later downsizes, the wires and substations do not vanish with it.

The question: does the customer who triggered the build pay for the build, or does everyone pay? If those costs go into the rate base, residential customers pay too. That is why "ratepayer protection" becomes the fight inside the fights. Utilities write special tariffs for large loads: long contracts, minimum payments, exit fees, security deposits. Simple intent: if the customer walks away, the customer still pays.

Some states add another layer: special review for very large loads. Once a project crosses into tens of megawatts, it stops being treated like an ordinary customer connection. It becomes a system event—not to block the project by default, but to slow it just enough to answer basic questions: what upgrades are required, what they cost, and who pays.

Policy is trying to catch up to scale. In Michigan, this tension surfaces during hearings at the Michigan Public Service Commission. Consumer advocates and environmental groups argue that residential customers deserve protection. The attorney general urges scrutiny. Meanwhile, the commission is asked to move quickly, because state officials are already celebrating the project as a win. Incentives bring deals to the table. The grid decides whether the deal can be served. Regulators decide whether the public will underwrite the risk.

The mechanics are dry but consequential. When a utility builds a substation for a data center, that cost enters the "rate base," the pool of capital on which the utility earns its regulated return. In Michigan, that return runs about ten percent. Ratepayers do not just pay for the substation. They pay for the substation, plus the utility's profit on the substation, forever. If the data center later downsizes or defaults, the infrastructure remains. The rates remain. Households absorb costs for a customer that no longer exists. This is why Attorney General Dana Nessel intervened in DTE's Saline application in late 2025, warning that "DTE customers" should not be "stuck footing the bill for a data center that never comes to fruition or uses far less electricity than projected." Consumer advocates call this a hidden subsidy—public risk for private returns, spread so thin across millions of bills that no individual notices, even as the system tilts.

9.7 Opacity

Transparency matters here too—or rather, its absence does. Incentive negotiations are often closed. Deal terms are confidential. Power contract provisions are redacted. Local officials and residents must accept the project without seeing the full economics.

In a small township, secrecy is a kind of power. It limits what citizens can challenge, because you cannot argue with numbers you are not allowed to see. Public hearings feel performative. People speak from intuition and fear, while the decisive terms live in private documents. A permanent asymmetry: developers negotiate with full knowledge of their own margins, while local officials bargain in the dark.

Developers argue they need secrecy to protect competitive information. Communities argue that if the public is carrying risk, the public should see the terms. Saline becomes a case study in what secrecy does to trust. Residents learn key facts late. They vote through their township board to reject the project. The developer sues. The township cannot afford years of litigation. It settles. The decision flips. Democratic voice is heard, and then it is outspent.

Accountability measures enter here. Some states add reporting requirements: disclose investment, disclose jobs, disclose timelines. Some add clawbacks: if promised jobs or investment do not materialize, incentives can be reduced or reclaimed. Imperfect tools—enforcement can be uneven, and agencies have incentives to declare success, not to hunt for failure. But the direction matters, because secrecy is not just a moral problem. It is a practical one. When deals are hidden, communities cannot compare notes. Regulators cannot learn. Bad terms repeat, because no one is allowed to see them clearly.

The demand for transparency is growing, pushed by watchdog groups, journalists, and local officials who have lived through the first wave and do not want to relive it blind. Whether that demand is producing results is another question. Disclosure requirements exist on paper, but enforcement is uneven, and developers still negotiate many terms behind closed doors. None of this is unique to Michigan. The same pattern plays out wherever data center incentives are offered: a state passes a package to attract investment, communities learn the details late, and the push for accountability comes after the deals are already signed. People are asking louder. The system has not yet answered.

9.8 What Incentives Buy

What do we do with all of this? We do not have to pretend incentives are pure evil to see the structure. Incentives pull private capital toward public subsidy. They shift risk onto public systems built to absorb it: utility rate bases, municipal budgets, state revenue. They create urgency, compress deliberation, and make local governance feel like a formality. Then they wrap all of that in the language of jobs.

In Saline, SB 237 matters. The race to the bottom matters. Ratepayer protections matter. Transparency matters. But incentives still do not change the core constraints we have tracked throughout. Power is still the constraint. The grid is still slow. Land is still contested. Incentives do not erase those facts. They just decide who pays to collide with them.

They decide something else too: where the tokens go. The student in Ann Arbor does not think about sales tax exemptions when she types a question into a chat box. But incentives shape the map that decides whether her request lands in Virginia, Texas, or a field outside Saline.

In the next chapter, we widen the lens again. Because this is not only a story about state legislatures and township boards. It is also a story about national strategy—about export controls and supply chains, about allied coordination and foreign capital, about what it means when a rural zoning vote becomes, like it or not, a piece of geopolitics.

CHAPTER TEN

The Geopolitics

On January 21, 2025, President Donald Trump stood in the Roosevelt Room beneath a screen that read "Stargate Project." Behind him, cabinet officials and Silicon Valley billionaires stood together in a single row—the kind of lineup that does not happen at ordinary press conferences. Reporters leaned forward in their seats. The tableau was strange, and it promised something large.

Trump announced Stargate, a venture that would invest five hundred billion dollars in AI infrastructure in the United States. Executives took turns at the microphone.

Larry Ellison, the founder of Oracle, a company that runs cloud computing infrastructure for corporations and governments worldwide, described data centers under construction in Abilene, Texas—buildings the size of shopping malls rising from the desert. Sam Altman, the chief executive of OpenAI, the company behind ChatGPT, called it a project of an era. Masayoshi Son, the head of SoftBank, a Japanese investment conglomerate that has poured tens of billions into technology companies, outlined the financing: deploy a hundred billion dollars quickly, then aim for five hundred billion within four years.

What Stargate actually is—a joint venture, a holding company, a loose coalition of private commitments—has never been fully clarified, and remains ambiguous more than a year later. But the number was real enough to fill the room, and the terms that emerged carried weight: ten gigawatts of computing power by 2029, foreign sovereign capital in the financing stack, NVIDIA chips, American cloud companies running the facilities. This was more than corporate ambition announced from a White House podium. It was national strategy. At that moment, "data centers" stopped sounding like real estate and started sounding like power.

10.1 The New Oil

For most of the twentieth century, great powers competed over strategic resources like oil, uranium, and semiconductors. These were not merely economic inputs. They were the raw materials of military capacity and geopolitical influence. A country that controlled them held power. A country that lacked them was vulnerable.

AI is becoming the strategic resource of the twenty-first century. We have known for years about military applications: surveillance, targeting, logistics, cyber operations, intelligence analysis. What changed in the 2020s is the recognition that AI capability depends on physical infrastructure. The transformer architecture revealed a strange pattern. When models grow larger and train with more compute, they do not just get better at the same tasks. They begin to do new things entirely. Capabilities emerge from scale. The race becomes a race for compute.

Compute, in practice, means data centers. Not the kind built to host websites and run enterprise applications. The kind built to run thousands of specialized AI chips, continuously, for months on end. Training a frontier AI model is not a single computer working hard. It is a vast cluster of thousands of GPUs running in parallel, day and night, for weeks or months, the training run humming like an industrial process. Training GPT-4, for example, reportedly required roughly twenty-five thousand GPUs running for months. A job like that requires a single facility large enough to house all of those chips and powerful enough to keep them running at once. Only a handful of places on Earth can do it.

A training run of that scale requires power, cooling, and a network fabric built for a machine that behaves like one enormous computer. When politicians and executives say "ten gigawatts by 2029," they are not speaking in metaphors. They are describing the physical limiters of capability. Frontier AI demands facilities that can feed and cool those chips: power, which means land, which means permits and politics. The stack becomes a national concern because the stack becomes a national bottleneck.

10.2 The Chokepoint

Compute is strategic. Compute is also fragile. Not because the math is hard, but because the supply chain is remarkably narrow.

Advanced AI chips are built using extreme ultraviolet lithography—machines that use light with wavelengths shorter than visible light to etch circuits smaller than a virus onto silicon. Only one company in the world makes these EUV machines: ASML, in the Netherlands. Those machines feed a small number of fabs capable of producing leading-edge chips: TSMC in Taiwan, Samsung in South Korea, and a limited set of facilities elsewhere. China lacks access to EUV equipment. The world's AI capability rests on a chokepoint that is partly Dutch, partly Taiwanese, partly American—and partly Japanese, German, and Korean, depending on which component you trace.

The EUV machine itself is a kind of international artifact, assembled from thousands of components that cross borders. Optics come from Germany. Precision metals come from Japan. Chemical systems come from multiple countries. Software and sensors come from the United States and Europe. No single country owns the whole process, which means no single country can control it alone.

The United States wants to keep these advanced manufacturing capabilities out of the hands of foreign adversaries—above all, China. But American law only reaches so far. If the Netherlands keeps selling lithography machines, or Japan keeps shipping precision manufacturing equipment, then restrictions imposed by Washington leak at the borders. Effective export controls require allied coordination: the Netherlands, Japan, South Korea, and Germany all agreeing to treat a commercial supply chain as strategic infrastructure and to restrict sales accordingly. That kind of coordination is difficult to build and even harder to maintain, but without it, the most advanced chipmaking tools eventually reach the countries the controls were designed to exclude.

The stakes are concrete. Taiwan manufactures the majority of the world's most advanced chips. Threaten Taiwan, and you threaten the supply chain that feeds every AI data center in America. If that chain breaks, the infrastructure we are building on American farmland sits waiting for hardware that may never arrive. Without chips, all the concrete in the world does not produce compute.

10.3 The China Challenge

The China challenge in its most basic form: China wants AI capability for the same reasons the United States wants it: economic power, military strength, control over information, national prestige. In response, the United States has treated advanced chips as dual-use technology. Dual-use means technology that works for both civilian and military purposes. The same AI that powers chatbots could power military targeting systems, making regulation nearly impossible. In October 2022, the Commerce Department issued export controls restricting the sale of advanced AI chips and chip-making equipment to China.

The goal is not just to keep specific chips out of specific hands. The goal is to slow an entire industrial trajectory. For most of the postwar era, American export controls targeted things that were obviously military: missile components, nuclear materials, weapons systems. The 2022 rules are different. They restrict categories of civilian technology because the same hardware that powers a chatbot can power a military targeting system. The United States decided that the risk of dual use was high enough to justify restricting commerce that would otherwise be perfectly legal.

China protested, searched for workarounds, and built alternatives. Huawei developed the Ascend line as a substitute for NVIDIA chips. Chinese firms stockpiled chips ahead of restrictions. Domestic foundries pushed older manufacturing techniques to their limits. In response, the allied coordination described above began to take shape—not in a press release but in quiet export licenses and negotiated restrictions across the Netherlands, Japan, and other partner countries. The point: keep the most advanced part of the supply chain out of China's reach for as long as possible.

From a U.S. perspective, every American data center built represents capacity China cannot easily match.

10.4 Policy in Flux

Then the policy itself started to shift. The Biden administration wrote the initial export controls in 2022 and tightened them in 2023 and 2024. In its final days, it proposed an "AI diffusion" rule that would have divided the world into tiers: close allies could buy advanced chips freely, countries in the middle faced quotas, and adversaries faced outright bans. The rule was scheduled to take effect in 2025. The incoming Trump administration began unwinding it before it could be enforced.

The instability matters far beyond Washington. A data center operator planning a campus with a decade-long horizon needs to know whether its customers will be allowed to buy chips, where those chips can legally operate, and whether foreign demand will be constrained or simply rerouted through intermediaries. These are questions worth billions of dollars, and the answers change with each administration. Policy uncertainty also shapes the politics of the buildout itself. When export controls are tight, domestic data centers look like national defense infrastructure. When controls loosen, the same buildings look more like ordi-

CHAPTER 10. THE GEOPOLITICS

nary commercial investments. The strategic framing shifts depending on which set of rules is in force.

In late 2025, the Trump administration approved the export of a powerful NVIDIA chip to China under a revenue-sharing structure, with the reasoning that American companies should capture the value from sales that would otherwise happen through smuggling networks and third-country intermediaries. Critics called it capitulation—handing advanced technology to an adversary. Supporters called it pragmatism—better to sell legally and collect revenue than to watch the chips move through back channels anyway.

That debate points to a deeper problem with enforcement. Smuggling is real. Intermediaries in third countries buy chips legally and resell them to restricted buyers. Paperwork is altered to disguise destinations. Demand for advanced chips is strong enough that buyers will pay a premium and accept the legal risk. No matter how carefully the rules are written, when a technology is scarce and valuable, it will move. Export controls can slow that flow and shape it, but they cannot stop it entirely.

The result is a dilemma with no clean answer. If chips leak through restrictions anyway, then strict rules cost American companies revenue without delivering the strategic benefit they were designed to produce. But if the government loosens the rules to capture that revenue, it speeds the very competitor it is trying to slow. Neither approach works perfectly. Congress, recognizing how large the stakes have become, has tried to pull more of this authority into legislative oversight rather than leaving it to executive discretion alone. The infrastructure buildout is happening inside this contest—shaped by constraints, incentives, and political cycles that shift with every election.

Stargate exists against this backdrop. When the president celebrates a massive investment in compute, he is not only promising jobs and economic development. He is declaring intent. The United States will build the capacity that defines the next generation of military and economic power.

The structure of Stargate reflects the logic of acceleration. Each partner fills a role that would otherwise be a bottleneck. OpenAI develops the AI models and consumes the compute. Oracle builds and operates the data centers. SoftBank, along with foreign sovereign capital, provides the financing to move at a pace that normal capital markets would not support. NVIDIA supplies the chips. The partnership exists because no single company can move fast enough on its own. The United States is trying to outrun its own constraints: interconnection queues, transmission upgrades, transformer supply chains, workforce training. Speed is a strategic choice, because in this competition, falling behind by a few years may mean falling behind permanently.

An awkward truth policymakers keep running into: policy optimized for selling chips is not always policy optimized for building domestic data centers. Loosen exports, and some AI workloads run abroad instead of on American campuses. That can reduce the urgency for domestic buildout even while it increases revenue for chip makers. Goals conflict. No one has figured out how to resolve the tension.

10.5 The Bloc

Allied coordination extends well beyond export controls. Energy security now matters. Europe, after the energy shocks of the early 2020s, knows what it means to run a modern economy with uncertain supply and high prices. Data centers are energy-hungry. AI data centers, more so. Expensive and scarce electricity means expensive and scarce compute.

The United States has an advantage: domestic natural gas, a large existing grid, a base of nuclear plants, fast growth in renewables. Power is not easy—interconnection queues stretch for years, transmission upgrades take decades, and grid capacity limits site selection. But the United

CHAPTER 10. THE GEOPOLITICS

States can plausibly build the energy needed for AI at a scale many other countries cannot match. Domestic production covers a large share of demand. Fuel supply is less exposed to geopolitical shocks than in regions that rely heavily on imports. No guarantee of cheap power, but a different strategic picture. A country that cannot reliably produce or procure electricity for compute cannot reliably produce capability. Energy becomes geopolitics. Compute follows power. Power follows fuel and policy.

Data sovereignty adds another layer. Countries care about where their data lives and where their compute happens. Some regulate this directly through explicit localization requirements. Others do it indirectly: procurement rules, privacy law, security standards. The effect is the same. The world fragments into jurisdictions, each with its own rules for where data centers can be built and what they must do. Global companies respond by building data centers in each region where they need to comply, turning legal requirements into construction projects.

Regulatory divergence amplifies that fragmentation. The European Union leans toward stricter requirements: energy efficiency reporting, deeper environmental review, more formal public consultation. The United States is more fragmented—utility commissions regulate power, local governments regulate land, federal agencies touch pieces of emissions, water, and grid planning without anyone holding the whole problem. China pushes in its own direction, sometimes steering data center development toward regions with certain energy characteristics and away from constrained cities. Developers ask not only "Where is power available?" but also "Where is permission available?" and "Where is legal risk lowest?" The regulatory map becomes part of the siting map.

Competition is not only between the United States and China. Europe tries to balance competitiveness with regulation. India courts investment while building its own domestic AI capability. The Middle East uses sovereign capital to buy a seat at the table: funding infrastructure, partnering with U.S. firms, positioning itself as an energy-and-capital hub. Smaller countries make their own bets, often shaped by a single factor: reliable power and a regulatory regime investors can live with.

The AI buildout is a new kind of development story. The industrial input is electricity. The export is capability. Geography is written in transmission maps, not shipping lanes.

10.6 Dual-Use Reality

Military and intelligence applications make everything harder. AI that summarizes a contract and AI that assists in targeting can share techniques and hardware. Data centers serving consumer chatbots can also serve defense work. Dual-use reality means infrastructure becomes a strategic target. A facility is not just a business asset. It is a concentration of national capability that foreign adversaries have reason to monitor, disrupt, or target.

Operators respond by hardening sites, adding redundancy, distributing capacity across geographies, building behind-the-meter power plants. The basic fact remains: concentrating critical compute in a handful of campuses creates risk. Adversaries know this. They plan around it. Dual-use reality also complicates regulation. If a facility is "just commercial," local opposition can be treated as a normal planning dispute. Frame it as strategic infrastructure, and pressure to override local concerns grows. That shift in framing can happen without any change in what the building looks like from the road.

There is also the question of workforce. A gigawatt-scale campus needs people who know how to operate complex systems—power distribution, cooling, fire suppression, network fabrics, security protocols.

Some of that talent has been built over decades in places like Northern Virginia. New regions entering the market have to build it from scratch.

Community colleges start technician programs. Companies run training academies. Unions adapt apprenticeships. Training takes time. For many roles, experience is the real credential. This experience gap creates a hidden advantage for established hubs. Northern Virginia is not just fiber and substations. It is people who have already lived through outages, thermal incidents, generator failures, security events. New regions have to build that muscle memory while the facilities are already humming—possible, but risky, with less slack in a system running at this scale.

Immigration policy intersects directly. Many of the engineers who run American infrastructure were born elsewhere. Tighten visas, and the pool narrows at the same moment demand is surging. Results show up in mundane ways: positions stay open for months, wages rise, operators fly people in from other regions to keep facilities running because the campus cannot wait for a local training program to graduate its first cohort. For new markets, this creates another tension: the community hosts the infrastructure, but the most specialized jobs commute in. Workforce development can close that gap over time, but training programs take years to produce experienced workers, and the buildout is not waiting. Money cannot instantly create people who have already lived through failures and learned how to prevent the next one. Workforce becomes a strategic resource too.

10.7 Local Meets Global

All of this connects back to Saline. Initiatives like Stargate create demand for data center sites far beyond Texas, and those projects need power, land, supportive state policy, and grid access. Michigan offers all of those: a utility that can serve a large load, land outside a major university town, and a state that has signaled it will compete through incentives.

10.7. LOCAL MEETS GLOBAL

When the governor announced the project, she framed it as having national significance, not only local prosperity. In that context, local opposition becomes harder to honor. Not because local concerns are invalid—they are real: electricity consumption, environmental impact, water use, traffic, noise, rural character. But they compete with considerations that do not fit inside township law: great power competition, export controls, military applications, national political agendas. When a township board says no, the conflict is no longer just between a developer and a community. It is between a community and a story the nation is telling about itself.

The new reality of AI infrastructure: decisions that once seemed purely local now carry implications that policymakers in Washington, Beijing, and Brussels track closely. The infrastructure enabling American AI capability has to be built somewhere. Communities that host it participate in the competition shaping this century, whether willingly or not.

The token's journey is not just a trip through fiber and transformers. It runs through relationships. A question typed in Michigan can be answered by chips designed in California, manufactured in Taiwan, installed in servers assembled in Texas. Foreign sovereign capital can finance the whole project. Export rules written in Washington and negotiated with allies overseas govern its operation. That web of dependencies decides where capability can exist. Geopolitical pressure feeds directly into design choices. Time is strategic, so we build faster. Power is strategic, so we build where the grid can serve us. The supply chain is strategic, so we change what we build and how we cool it.

Nations have always built the infrastructure they believed their futures depended on: canals, railroads, telegraph lines, power grids, highways, satellite networks. Each generation's strategic buildout reshaped the landscape and the communities that lived on it. This one is no different, except in speed and scale. The question is whether the current model—gigawatt campuses on rural land, powered by gas, financed by global capital, justified by great power competition—is the only way to build it. In the next chapter, we look at what could change.

CHAPTER ELEVEN

What Could Change

In a basement office at a Lansing utility building, an engineer pulls up a schematic on her monitor. The diagram shows something that would have seemed strange a few years ago: heat flowing backward. Instead of a power plant sending energy to customers, a data center will send waste heat into the city's district heating network. District heating uses underground pipes to carry heat from central sources to buildings throughout a city—common in Europe, rare in the United States. Ten miles of those pipes will carry warmth to office buildings downtown.

This is the plan for Deep Green's proposed facility—a twenty-four-megawatt data center on a vacant parking lot, announced two months ago and expected to break ground this spring. If it works as designed, it will offset a quarter of the natural gas the city burns for heating. No farmland converted. No township transformed. A data center that solves a city's problem rather than creating one.

We have traced a token backward: inference to silicon, data centers to power, grid to generation, land to money, incentives to geopolitics. That tour explains why the default answer today is a gigawatt campus on rural land. But defaults can change. Cheaper compute, different building patterns, rules that reward integration over speed—any one of them could shift the equation.

11.1 The Efficiency Revolution

The first force that could reshape the buildout is efficiency. We saw in Chapter 2 how DeepSeek's January 2025 release sent shockwaves through the industry—a Chinese lab matching top American AI systems at a fraction of the reported cost, then freely distributing the model as open weights for anyone to use. NVIDIA lost seventeen percent in a single day. The architectural innovations were real, and the rest of the industry quickly adopted them. Inference costs dropped by a factor of a thousand over three years, by some estimates.

For anyone watching the data center buildout, this raised an obvious question: if AI gets dramatically more efficient, do we still need all those gigawatt campuses?

11.2 The Edge

Efficiency has a geographic dimension. Not all AI computation needs hyperscale facilities. Edge computing means running AI on local devices—phones, laptops, nearby servers—instead of sending everything to distant data centers, resulting in faster response for users and less strain on centralized infrastructure.

Apple runs many AI requests entirely on-device. Neural engines handle tens of trillions of operations per second. Google's Gemini Nano processes requests on phones without ever touching the cloud. Laptops ship with dedicated neural processing units—high-end chips reach eighty-plus trillion operations per second, enough for useful local inference.

Self-hosted inference is no longer exotic. Tools like vLLM and Ollama let companies run capable models on their own hardware. Small language models—three billion parameters, sometimes less—handle tasks that once required frontier systems. By 2027, analysts predict organizations will use small, task-specific models three times more often than general-purpose large ones.

A hybrid architecture emerges. Hyperscale facilities still handle training and batch processing. But inference—answering questions, generating images, running assistants—splits between centralized and distributed locations. Some stays in the cloud. Some moves to regional data centers closer to users. Some runs on the device in your hand.

For communities and investors, the question is how much that split changes the infrastructure equation. If half of inference moves to the edge, does that mean half as many data centers? Does it change which places get built?

11.3 The Jevons Trap

History offers a warning. In 1865, British economist William Stanley Jevons observed that steam engines were burning less coal per unit of work than ever before. But the country was burning more coal in total, not less. The reason was straightforward: cheaper operation made steam power affordable for factories, mines, and mills that previously could not justify the cost. New users more than offset the savings from each individual engine. Efficiency did not reduce demand. It expanded the market.

Computing followed the same pattern for decades. Moore's Law—the observation that the number of transistors on a chip roughly doubles every two years, driving costs down—drove down the cost per transistor by orders of magnitude. We did not get less computing. We got computing everywhere. LED bulbs use eighty percent less energy than incandescents, yet we got more lights, longer hours, projects like the Vegas Sphere. Cloud computing made servers dramatically cheaper to operate. Enterprises responded by spending more on IT, not less.

AI efficiency is already showing signs of the same dynamic. After DeepSeek's January 2025 release, panic gave way to something else. Reasoning costs dropped ninety percent. Usage exploded. By October 2025,

NVIDIA had recovered fully, reaching a five-trillion-dollar market cap. The company supposedly disrupted by efficiency became its biggest beneficiary.

Microsoft CEO Satya Nadella offered his shareholders a reassuring frame on social media after the DeepSeek announcement: "Jevons paradox strikes again."

The mechanism is straightforward. Cheaper AI makes previously uneconomical applications viable. Medical diagnosis at scale. Personalized tutoring. Autonomous agents running continuously instead of responding to queries. Reasoning models that think for minutes instead of milliseconds, using a hundred times more compute per answer. Each efficiency gain creates new demand that can exceed the savings.

Efficiency is not irrelevant. But it changes *who* can afford AI and *what* they can do with it—without necessarily reducing total infrastructure. Gigawatt campuses may still get built. They may just serve a broader set of applications than anyone initially imagined.

11.4 THE BROWNFIELD ALTERNATIVE

Even if total demand stays high, location is not fixed. Converting farmland into data centers, cutting transmission corridors through forests and wetlands—that model dominates today. But it is not the only option.

Deep Green's Lansing project, described at the opening of this chapter, is one template. It is not unique. Switch converted the former Steelcase corporate campus in Grand Rapids into a major colocation facility. xAI transformed an abandoned Electrolux factory in Memphis into its Colossus supercomputer in one hundred twenty-two days. Google repurposed an old Finnish paper mill, using the same seawater pipes that once cooled turbines to now cool servers.

A pattern emerges: brownfield works when economics align. Switch got exceptional existing infrastructure at a fraction of new-build cost. xAI needed speed more than optimization and found a warehouse that was good enough. Google found a site where natural cooling solved its biggest operational challenge.

Barriers remain. Contaminated sites require environmental assessment and remediation that can cost millions and take years. Old industrial buildings rarely have the floor capacity and structural strength that dense server racks demand. Urban power grids are built for dispersed residential load, not concentrated industrial draw. Timeline uncertainty kills deals. Hyperscalers and their lenders want predictable schedules. Discovering unknown contamination mid-excavation can delay a project indefinitely.

The cost gap is real. For a thirty-megawatt facility, brownfield development might add twenty-five million dollars in upfront costs compared to greenfield—demolition, remediation, and retrofit. On a project that might cost three hundred million dollars total, that gap is enough to push expected returns below the threshold that lenders and investors require. When incentives are equal, developers choose the lowest-cost, fastest site—almost always a cornfield.

The counterargument is scale. Fifty facilities in the twenty-to-thirty-megawatt range add up to about a gigawatt—the same capacity as one massive rural campus, spread across dozens of cities that already have infrastructure, workers, and heating demand.

11.5 Building Faster

Developers are also trying to change how fast they can build the box. Factory-built electrical rooms, prefabricated cooling skids, standardized data halls—construction becomes assembly. When it works, a campus adds capacity in months instead of years.

Modular construction does not outrun the grid. In many regions, interconnection and transmission upgrades still take longer than building. Some operators chase behind-the-meter power or finance their own infrastructure, moving the bottleneck back to permits and steel.

CHAPTER 11. WHAT COULD CHANGE

Speed changes the power dynamic in negotiations. A developer who can break ground in weeks has leverage over a township that needs months to review plans, consult residents, and assess infrastructure impact. The leverage compounds when multiple localities are competing for the same project. Developers can—and do—play sites against each other: approve our plan on this timeline, or we take the investment to the next county. Local officials face a choice between rushing approval with minimal diligence or watching a rival community claim the jobs and tax revenue. The result is a race to the bottom. Townships waive review periods, accept vague commitments on infrastructure costs, and lock in tax abatements they may later regret—all because slowing down feels like losing.

That dynamic is precisely why disclosure requirements, mandatory review periods, and clear rules about who pays for infrastructure upgrades matter more as construction accelerates, not less. State and federal standards can break the race by setting a floor that no locality can undercut. Without them, speed favors the developer at the expense of every community at the table.

11.6 What Policy Could Change

Greenfield preference is not a law of nature. It is a policy outcome. Policy can change.

Germany's 2023 Energy Efficiency Act requires new data centers to reuse progressively higher shares of waste heat: ten percent for facilities commissioned after July 2026, fifteen percent after July 2027, and twenty percent after July 2028. Denmark removed a tax that had made waste heat recovery uneconomical, reviving stalled projects. Stockholm pays data centers for their waste heat through predictable long-term contracts. More than thirty data centers now feed the city's district heating network, warming tens of thousands of apartments.

11.6. WHAT POLICY COULD CHANGE

In July 2025, a federal executive order directed the EPA to identify brownfield and Superfund sites suitable for data center development, expediting environmental reviews and providing financial support for qualifying projects. Michigan extended tax exemptions for brownfield data centers by fifteen years beyond the standard greenfield term—though repeal efforts began within a year of passage.

As we saw in Chapters 5 and 7, brownfields lose to cornfields for structural reasons: contamination, grid topology, title complexity, and building inadequacy. Those barriers are real, but they are policy choices, not laws of nature. Public funding for remediation would close the upfront cost gap. Expedited grid interconnection for sites with existing infrastructure would reduce the timeline penalty. Streamlined title review would cut the legal delays that make lenders walk away.

Policy could also work from the other direction. Waste heat mandates—like those Germany and Denmark already enforce—would give urban sites near buildings that need heating an advantage that remote farmland cannot match. The cornfield is the default today not because it is the best answer, but because policy has not asked developers to consider any other one.

None of this would eliminate greenfield development. Hyperscale training facilities still want hundreds of megawatts in a single location, and few brownfield sites can deliver that. But the buildout is not monolithic. Smaller inference facilities, regional edge deployments, enterprise colocation—all could shift toward urban and brownfield sites if policy made that path competitive.

Nordic countries demonstrate what is possible. Finland expects data center waste heat to supply sixty-five percent of district heating in some regions once current projects are operational. That heat replaces natural gas, cuts carbon, and reduces energy costs for residents. Data centers become assets to cities, not just industrial neighbors.

11.7 THE COOLING FRONTIER

The liquid cooling technologies described in Chapter 3—direct-to-chip and immersion systems—do more than solve a thermal problem. They change the siting calculus. Denser facilities need less land for the same compute capacity. Liquid systems capture heat at higher temperatures, which makes waste heat recovery more practical. A facility designed for immersion cooling and heat recovery fits the urban model better than one designed for air cooling and cooling towers.

Adoption is slow. Operators who sell uptime are cautious about methods with fewer decades of track record. Immersion changes maintenance, repairs, and supply chains. But the direction is clear, and the question for siting is when, not whether.

11.8 WHO MOVES FIRST

The economics of innovation shape who moves first. Hyperscalers spend enormous sums on research and development. They treat the data center as a product they are constantly improving. They can afford to test immersion systems, new power distribution schemes, new networking approaches—then deploy what works across their fleets.

Smaller operators often wait. They depend on vendors to commercialize what hyperscalers prove out, or on hyperscalers to standardize a design enough that lenders and insurers feel comfortable. A gap emerges: the biggest players move first, the rest follow.

Even proven designs deploy unevenly. New facilities are easier to build with new ideas. Existing facilities are harder to retrofit. A vintage effect emerges. The newest campuses tend to be more efficient and more capable—not because operators are smarter, but because they are building with a newer toolkit.

For places like Saline, this matters. A facility is a huge investment up front, but not one-and-done. Over its lifetime, equipment will be replaced many times. Cooling systems upgraded. Power distribution reworked. If innovation is incremental, the facility evolves smoothly. If innovation is abrupt, the facility can be stranded faster than anyone planned.

11.9 Which Future

This chapter has traced several forces that could change the trajectory of the buildout. Efficiency gains could reduce the compute needed per task—but history suggests they expand demand rather than shrink it. Edge computing could distribute inference away from centralized campuses—but training and heavy workloads still need scale. Brownfield development could steer projects toward cities instead of farmland—but the cost and timeline penalties remain real without policy intervention. Cooling technology could make smaller, denser, urban facilities practical—but adoption is slow and operators are cautious. Modular construction could accelerate the buildout—but speed gives developers leverage over communities competing for investment, and faster timelines make oversight harder, not easier. And innovation itself is unevenly distributed: hyperscalers adopt new designs first, smaller operators follow years later, and facilities built today risk being stranded if the technology shifts abruptly.

None of these forces, on their own, overturns the current default. Gigawatt campuses on greenfield land dominate because they are fast, cheap, and proven. That will not change unless policy changes the incentives. Waste heat mandates, brownfield remediation credits, carbon pricing, interconnection reform, farmland conversion costs, minimum review periods, state and federal standards that prevent localities from undercutting each other—the tools exist. They have not been used.

Concrete is not yet poured on most announced facilities. The decisions being made now—by developers choosing sites, by officials approving permits, by legislators writing incentive packages—will determine whether this buildout repeats the same pattern in every state or adapts to the alternatives this chapter describes. Those choices remain open. But they will not remain open indefinitely. The next chapter asks what hap-

CHAPTER 11. WHAT COULD CHANGE

pens next: not one predicted future, but several possible ones, and what it would take to be ready for each.

CHAPTER TWELVE

Our Future

Monday, January 14, 2030. The sun is not yet up. A brand-new school bus idles at the corner of Michigan Avenue, its harsh blue LED headlights cutting the dark.

Frank squints past it, one hand on the wheel, the other around a paper cup of gas station coffee. Across the road, the substation fence glitters with frost. Beyond it, the data halls sit like long gray ships, surfaces catching the weak winter light. A low hum leaks into the cold air, steady enough to fade from notice.

Four years have passed since the first construction equipment arrived in Saline Township. What once felt impossible is now routine. Trucks have their own rhythm. The horizon has new shapes. Around the clock, the facility draws power, its cooling towers exhaling pale plumes into the January sky.

That scene is a projection, not a prediction. We do not know what Saline will look like in 2030. That is the honest answer. AI infrastructure depends on variables no one can predict with confidence. Capabilities can speed up or slow down. Demand can overshoot or fall short. Policy can tighten or loosen. A single breakthrough can change what computing requires overnight.

Instead of pretending we can forecast one correct future, we do something more useful: we plan in scenarios, not to predict, but to prepare.

We started with a student typing a question into a chat window. We followed that question backward—from the token on her screen to the chips that processed it, the cooling systems that kept those chips running, the grid and generation sources that delivered power, the land deals and incentive packages that put the building there, and the geopolitical rivalries that shaped who got the chips in the first place. Once you see the full stack behind a single query, the future stops looking like a single forecast. It becomes a set of choices under constraints.

12.1 Planning in Uncertainty

Sarah thinks about scenarios every day. She works in a regional grid control room, where operators cannot afford to plan for a single future. During a break between shift handoffs, she explains their discipline: "We run contingencies constantly. What if a generator trips? What if demand spikes? What if a transmission line goes down?" The grid works because operators think through alternatives before the crisis arrives. They rehearse failure so they can prevent it.

That discipline applies here. Data center load growth is the biggest variable in many regional forecasts. If developers build what they announce, the grid needs massive new transmission. If they stall or scale back, utilities overbuild. Ratepayers then cover infrastructure that never needed to exist.

Sarah does not get to choose between those outcomes. She has to keep the grid running no matter which one arrives.

12.2 Three Scenarios

No one knows which future we are headed toward. But for this purpose, we will narrow the range to three. In the first—Boom—demand keeps compounding and the buildout accelerates. In the second—Bust—demand disappoints and projects stall. In the third—Scatter—efficiency gains and policy changes spread computing across many smaller sites instead of a few massive campuses. None is a prediction. The point is to ask: what holds true in all three, and what should we do now regardless of which one arrives?

Boom

In this future, demand keeps compounding. AI becomes a normal layer in business software, search, customer support, medicine, logistics, education. Every product adds an assistant. Every company runs models internally. Efficiency gains lower costs, and lower costs expand usage faster than efficiency shrinks the footprint.

The infrastructure race continues. Projects we have described no longer look like overreach—they look like the minimum. Grid planners stop treating data center load as an edge case. It becomes the main story. More natural gas plants get built because they can be permitted faster than renewables or nuclear. Nuclear projects move forward, but slowly—too slowly to meet near-term demand. The result is that carbon emissions from the power sector rise in the short term, even as companies promise to reach net zero in the long term. Direct costs climb—electricity rates, land prices, construction labor. Indirect costs follow—road wear, water stress, strained local services. And the externalities that no one pays for directly—carbon, noise, lost farmland—grow with every new facility.

In Saline, the campus expands through phases, pushing beyond the original plan as demand keeps climbing. Jobs rise too—not into the tens of thousands, but into the hundreds. A township that once expected a few

CHAPTER 12. OUR FUTURE

staff members to handle zoning now needs ongoing oversight, emergency planning, constant coordination.

Economically, it looks like a win: construction work is steady, property tax revenue grows, some local residents get trained and hired into operations roles. But the impacts intensify too. More traffic. More industrial presence. More political fights about water, noise, and grid stress.

In this world, stranded assets—infrastructure built for demand that never comes—are not the risk. The risk is political. The buildout moves faster than communities can absorb it. Residents who were never consulted watch their landscape change. Ratepayers who never agreed to higher bills see electricity costs rise. Local officials who fast-tracked approvals face voters who feel they were not asked. When people believe the process was fair, they accept outcomes they do not love. When they believe it was not, they fight—at the ballot box, in court, in township meetings that run past midnight. Even in the boom scenario, the buildout succeeds only if the public believes it was done legitimately. That trust is harder to build than a data center, and easier to lose.

Bust

In this future, demand disappoints. Maybe enterprise adoption is slower than the hype suggests. Maybe competition compresses margins and makes AI services less profitable. Maybe efficiency improves fast enough that the same capability can be delivered with far less infrastructure. The result: projects announced in the boom years exceed what the market needs.

Construction starts, then pauses. Campuses sit half-built. Some operators consolidate. Some projects change owners. Refinancing becomes the quiet killer: deals that looked fine at low interest rates turn fragile when debt rolls over at higher costs.

In Saline, the buildout is uneven. One or two phases complete. The site might be capable of drawing 1.4 gigawatts at maximum, but it rarely runs there because demand never materializes. Permanent jobs come in

12.2. THREE SCENARIOS

below projections. The tax base rises, but not enough to feel like a miracle.

The site is still transformed. Land is still gone. Barns do not come back. But the full promised scale never arrives. The community faces the hardest version of the question: the farmland is gone, the tax revenue fell short, and the jobs never reached the numbers that justified the project. Should the township have demanded stronger guarantees—clawback provisions, minimum employment commitments, infrastructure cost-sharing—before approving the deal? And if this deal was not worth it, was any deal?

This is why ratepayer protection and incentive accountability matter. If overbuilding is possible, public systems need ways to avoid eating the downside alone.

A starker version of this scenario haunts some observers: what if the transformative capabilities never arrive? Not a plateau, but a reckoning. Serious AI researchers, not just critics but working scientists, question whether current architectures can deliver the general intelligence being promised. Pattern-matching, however impressive, may not be understanding. If that skepticism proves correct, we are not building infrastructure for a plateau. We are building infrastructure for parlor tricks at industrial scale. The counterargument: even narrow AI generates real economic value—customer service, code assistance, drug discovery. The buildout may be justified even without breakthrough capabilities. But the honest answer is that no one knows which side is right. We are betting a trillion dollars on an assumption we cannot test until the facilities are built.

Scatter

In this future, technical and policy changes reshape the geography of computing. Efficiency breakthroughs make smaller facilities viable for more workloads. Edge computing absorbs a meaningful share of inference. Carbon pricing and waste heat mandates tilt economics toward urban brownfield sites. Hyperscale greenfield does not disappear. It just stops being the only game.

A sharper version of this shift is a discontinuity. A breakthrough in model design or chip architecture cuts compute needs by an order of magnitude. Then the question changes. Not where to build next, but what to do with what has already been built: half-used campuses, unfinished phases, infrastructure sized for a load that never arrives.

New facilities look different. Some are smaller regional inference centers: tens of megawatts instead of hundreds, located closer to users, integrated into urban infrastructure. Some are brownfield conversions—old factories and warehouses repurposed for compute. Others export waste heat from day one, turning cooling loads into heating assets.

In Saline, the campus might never reach its maximum planned load. It becomes one node in a distributed network rather than a singular monument. The facility still matters, still employs people, still pays taxes. But the story of Saline as a flagship gigawatt campus fades. The facility becomes one regional hub among many, not the defining symbol of the AI buildout.

This scenario does not eliminate the tensions we have traced. It redistributes them. Urban sites face their own politics: noise, traffic, neighborhood opposition. Brownfield development still struggles with contamination and timeline risk. The grid still needs upgrades, just in different places. But the concentration of impact shifts. Instead of a few rural communities hosting gigawatt campuses, many places host smaller facilities with different cost-benefit profiles.

12.3 What Holds

Across all scenarios, some truths hold.

Power remains the binding constraint. Even in a slowdown, electricity is not free. The grid stays slow relative to capital. Transmission takes years because it has to; equipment takes time because factories have limits.

Land, once transformed, does not reverse on a business cycle. A township that paves farmland for an underperforming campus does not get its farms back as compensation.

Money moves faster than institutions. Incentives and financing terms shift in months; transmission takes years. Geopolitics can reframe a local vote as strategic infrastructure overnight.

Legitimacy depends on process. Communities accept rapid change when they feel they chose it with clear information and fair terms. They resist when change feels imposed under pressure and secrecy. Legitimacy is not a nice extra. It is infrastructure too.

12.4 What to Watch

What should we watch? Start with physical indicators. Utilization rates tell us whether new facilities actually draw the power they were built for. Interconnection queues reveal whether utilities still see new requests or whether demand has slowed. Capital terms show whether money is getting cheaper or lenders are tightening.

Watch the political signals. Do regulators tighten cost responsibility rules—deciding who pays for grid upgrades when a large customer arrives? Do states expand incentives or face backlash and pull back? Does federal policy stay hands-off, or does it set minimum standards for water, carbon, and community process? Do export controls tighten, loosen, or oscillate? Do communities start winning more fights, or do lawsuits keep flipping votes?

Not every community accepts. In 2024, voters in Prince William County, Virginia—at the heart of Data Center Alley—rejected a zoning change that would have allowed a new data center campus near residential neighborhoods. The developer walked away. In Oregon, Jackson County commissioners denied a permit after residents raised concerns

CHAPTER 12. OUR FUTURE

about water and noise. In Wisconsin, a township moratorium gave a rural community time to study impacts before deciding. These are exceptions, not the rule. Most proposed projects proceed. But communities that push back are getting better at it. Townships share model ordinances and legal strategies. Noise and water protections drafted in one county show up in the next. The question for policymakers: should refusal be easier, or should the goal be shaping projects so they are worth accepting? Both paths exist. Neither is automatic.

Wild cards can blow up any baseline. A major data center incident triggering regulatory shock. A financial crisis freezing capital even when demand is strong. A grid emergency turning public opinion against large loads. A geopolitical disruption threatening chip supply chains. A security event that makes the world see data centers as targets, not buildings.

12.5 Policy Levers

Policy cannot control which scenario arrives. But it can shape how the arrival is managed. Other democracies are already experimenting with tools the United States has barely tried.

As we saw in the previous chapter, some countries have already built what American policy only talks about. Nordic cities pipe data center waste heat into district heating networks. Germany's 2023 energy efficiency law mandates it. These are not pilots—they are legal requirements that reshape where and how facilities get built.

Now consider the warnings. Ireland welcomed data centers with favorable tax treatment and fast approvals. The facilities came. By 2024, data centers consumed more than a fifth of Irish electricity—and the share keeps climbing. The government has imposed restrictions on new development. What looked like economic development became a constraint on everything else: housing, manufacturing, the grid itself. Northern Virginia shows signs of the same pattern. Loudoun County hosts the densest concentration of data centers on Earth, and the region's grid is strained to the point where new projects face years of delay. The lesson is not that data centers are bad. Concentration without planning creates problems that outlast the boom.

Singapore offers a third model. The city-state treats data center capacity as national infrastructure, allocating power through a managed process rather than a first-come-first-served scramble. Operators compete on efficiency and integration, not just speed. This approach requires more government coordination than American traditions favor. But it avoids the whiplash of boom and restriction.

These international examples point toward a menu of policy tools that American jurisdictions have underused.

Grid planning is the foundation. Transmission planning that assumes flat demand is no longer credible. If we want AI infrastructure, we need grid infrastructure to match—and we need to build it faster than the old pace allowed.

Waste heat mandates can turn cooling loads into heating assets. Requiring facilities to capture and reuse heat—or pay the cost of not doing so—changes the economics of urban integration. A data center that warms a neighborhood is a different neighbor than one that just hums.

Brownfield incentives can shift development toward sites already disturbed. Tax preferences, expedited review, and remediation support can close the cost gap between a contaminated industrial site and a clean cornfield. That gap will not close on its own.

Cumulative impact reviews can catch regional saturation before it becomes a crisis. Project-by-project review misses what happens when dozens of projects land in the same grid region. Triggers forcing broader assessment once a region crosses a threshold can prevent the next Loudoun County.

Water standards can limit consumption and require closed-loop cooling in stressed regions. The technology exists. The question is whether policy requires it.

Ratepayer protections can ensure that infrastructure costs fall on beneficiaries, not on households that receive no benefit. When a utility builds a substation to serve a data center, policy decides whether the data center pays or whether the cost spreads across everyone's bill.

Incentive accountability can tie public benefits to public outcomes. Publish terms. Require reporting. Build in clawbacks when promises are missed. End the practice of giving away decades of tax revenue on the promise that jobs and investment will follow.

Technical assistance can level the playing field. Small governments facing sophisticated developers need help reading thousand-page engineering documents and evaluating claims they have never encountered. States can provide that help. Or they can leave townships to negotiate blind.

Minimum review periods can take urgency off the table as a negotiating tactic. If policy sets a floor on evaluation time, developers cannot stampede approvals by threatening to leave for the next county.

None of these tools are magic. They do not guarantee good outcomes. But they give communities and policymakers tools they currently lack—the ability to set terms, demand transparency, and say no without losing every future opportunity.

12.6 The Question of Purpose

Policy is not just about where data centers go and who pays. It is also about what they are for. A democratic society transforming its land at this scale has a right to ask whether the transformation serves purposes worth the cost. That question is currently unanswerable, because the process does not require anyone to answer it.

Consider the national security framing. Some AI infrastructure genuinely serves strategic purposes: training frontier models that maintain technological advantage, supporting defense applications, enabling capabilities adversaries cannot match. When a project serves those purposes,

12.6. THE QUESTION OF PURPOSE

expedited approval and public investment may be justified. The nation has an interest in moving fast.

But national security has become a rhetorical umbrella covering far more than it should. Projects with no meaningful defense application invoke strategic competition to short-circuit review. Every data center becomes "critical infrastructure." Every delay becomes a gift to China. The framing works because it contains a grain of truth—and because communities have no way to evaluate which projects actually serve the purposes claimed.

The result is a kind of policy capture—when special interests use legitimate concerns to bypass normal oversight. Legitimate security concerns justify a fast-track process. That process then gets used for projects with nothing to do with security. Communities accept transformation in the name of national interest, without any way to verify that the national interest is actually at stake.

Behind all of this lies a deeper question that almost no one asks in public: what will the compute actually do?

Not all uses of AI are equivalent. Models that cure cancer are not the same as models that generate synthetic pornography. Models that improve weather prediction and disaster response are not the same as models that summarize the endless drivel of middle management memos. Models that support scientific research are not the same as models that target advertising or generate addictive short-form video content.

These are not abstract distinctions. They are choices about what a society builds, what it sacrifices, and what it gets in return.

Right now, the approval process does not engage with any of them. Developers are not required to disclose what workloads their facilities will run. Communities cannot condition approval on purpose. The operating assumption is that compute is compute—that a megawatt serving medical research and a megawatt serving algorithmic spam are equivalent, and that neither the community nor the public has any standing to distinguish between them.

That assumption deserves scrutiny. If we transform farmland, strain grids, consume water, and reshape communities, purpose should matter. Not every use of AI compute carries equal weight. A township asked to host a gigawatt campus might reasonably accept disruption if the facility trains models that cure cancer or improve climate prediction. The same township might reasonably refuse if the facility generates content corroding public trust or serves no purpose beyond advertising optimization.

This is not an argument for government control over what AI systems do. It is an argument for transparency and democratic choice. Communities should know, in broad terms, what they are being asked to host. They should be able to condition approval on commitments about use. They should have recourse when a facility's actual operations diverge from what was promised.

Policy could require disclosure of general use categories—not proprietary model details, but broad purposes. Training versus inference. Research versus commercial. Defense versus consumer. The categories need not be perfect to be useful. They just need to give communities something more than "trust us."

Policy could tie expedited approval to demonstrated public benefit. Projects genuinely serving national security or scientific research could move faster. Purely commercial projects could proceed through normal review. The distinction would force developers to justify the treatment they seek.

Policy could create accountability when purposes change. A facility approved for medical research that pivots to advertising should face consequences. The community accepted one bargain. Changing the terms should require renegotiation, not just a press release.

These tools assume institutions capable of wielding them. That assumption deserves scrutiny. Today, no single agency governs AI infrastructure end-to-end. Federal regulators handle environmental review, export controls, grid reliability. They act separately, often in tension. State utility commissions approve rate cases. Local boards approve zoning. No one asks whether the cumulative buildout serves the public interest, because no one has jurisdiction over the cumulative buildout. Regulatory capture is a risk at every level: the industry is sophisticated, well-funded, and focused; the agencies are fragmented, under-resourced, and reactive. Building the policy tools without building the institutional capacity to use them is like designing a car without training drivers. The governance deficit is as real as the infrastructure deficit—and harder to see.

None of this will happen automatically. The current system benefits developers who prefer opacity and speed. Changing it requires political will from communities usually outmatched and policymakers facing intense lobbying. But the question remains, whether or not policy addresses it: what is all this infrastructure for? Is the answer good enough to justify what we are giving up?

12.7 Keep Options Open

If we cannot know which future arrives, what can we do? Preserve options.

For communities, preserving options means phasing. Approve the first stage and make later stages conditional on clear triggers: demonstrated demand, verified impacts, enforceable mitigation, updated public review. Write agreements that do not assume best-case utilization. If a campus draws less power than planned, the community should still get water monitoring, road repairs, noise limits, financial commitments. If a campus expands, the community should get a real chance to renegotiate rather than being told, "You already said yes."

For policymakers, preserving options means building institutions that learn. Create technical assistance so local governments are not reading thousand-page documents alone. Require disclosures letting the public verify claims. Protect ratepayers with cost rules matching modern load scales. Treat grid planning like scenario planning, not like a single forecast.

For developers and investors, preserving options means designing for flexibility. Avoid single-purpose designs that cannot be repurposed if the market shifts. Be honest about timelines and risks. Do not treat every community as an obstacle to outlast.

For citizens, preserving options means participation. People who show up in township rooms, who ask for terms in plain language, who insist on disclosures and accountability—they are not slowing progress for sport. They are doing governance.

12.8 Learning Faster

Policy has to learn faster than it has. The lag between approval and outcome spans years. If we only learn from anecdotes, we will repeat the same mistakes across hundreds of projects.

Monitoring matters. Track promised jobs against actual jobs, tax revenue against public costs, power use against rate impacts, water consumption against local environmental outcomes. These are not academic metrics. They are feedback. Without feedback, institutions cannot adapt.

The Ohio township supervisor who gets a phone call next year will face a choice. Whether she has better tools than Saline did depends on choices being made today.

———

What would it mean for this buildout to go well? Not maximizing investment or minimizing opposition—those are metrics for particular interests. But arriving at 2030 with institutions intact, communities that feel they chose their path, a grid that works, an industry that earns its social license rather than extracting it.

That outcome is not guaranteed by any scenario. It has to be built. The same urgency driving the data center buildout should drive the governance buildout accompanying it. We are deciding, right now, what kind of infrastructure becomes permanent. We are deciding who bears costs and who captures benefits—and for how long. We are deciding whether places asked to host this transformation have a real voice or just a ceremonial one.

Those decisions are being made in utility commission hearings, township board meetings, state legislature sessions, corporate boardrooms. They are being made by people who may not realize they are making them. The window for shaping this buildout is open. It will not stay open forever.

The question of purpose has a labor dimension the approval process ignores. AI is not just a tool for answering questions. It is an automation technology, one that its creators explicitly design to replace human cognitive work. The student in the coffee shop gets her answer faster. Somewhere else, a customer service representative, a paralegal, a junior copywriter loses hours, then a job. Those displaced workers do not appear in township hearings. They have no standing to object. Yet the infrastructure we are building will reshape labor markets in ways we are only beginning to understand. If AI displaces millions of workers over the next decade—as some economists predict—the social contract implied by the buildout depends on policies that do not yet exist: retraining that works, safety nets that catch, transitions that feel chosen rather than imposed. We are building the infrastructure first and hoping the institutions follow.

This chapter has asked what could happen and what we should do about it. We traced three scenarios. We named the policy tools that exist but remain unused. We asked the hardest question—what all this compute is actually for—and found that no one in the approval process is required to answer it. We asked who bears the cost when AI automates the very

CHAPTER 12. OUR FUTURE

workers whose communities host the infrastructure, and found that no one in the approval process is required to answer that either.

But scenarios and policy levers are abstractions. The buildout is not. It is concrete and copper and displaced soil. It is a farmer who sold his land and did not live to see what replaced it. It is a township official who cast a vote he still thinks about every morning. It is a daughter whose job is being hollowed out by the same technology her father's vote helped bring to town.

We return now to Saline, to the people who live with the consequences.

EPILOGUE
The Token, Revisited

January 2030

FRANK left the township board in 2028. He cited his grandchildren when he stepped down—*Lily* is twelve now, *Marcus* ten, and they still ask him about the big computer—and a desire to spend winters somewhere warmer. But he has not moved yet. Something keeps him here. His wife says it is stubbornness. He thinks it might be something else. Witness, maybe. The need to see what he helped create.

His daughter *Beth* works in Troy now, in mortgage processing for a regional bank. She handles the files the AI flags as exceptions—the applications with irregular income, the self-employed borrowers whose tax returns do not fit the templates, the refinances where the appraisal comes in wrong. The algorithms process a thousand files for every one she touches. She is always on call. Her phone buzzes at dinner, at *Marcus*'s soccer games, at 2 a.m. when a closing is scheduled for the next morning and something has triggered a fraud alert. The software never sleeps, and someone has to approve the decisions it cannot make alone. Her salary has not risen in three years, though her hours have. She tells *Frank* that she is lucky to have the job at all. He notices she looks tired.

He thinks about that vote every day. Not whether it was right or wrong—he has made his peace with that. He thinks about whether they could have done it differently. Whether there was ever really a choice.

Epilogue

He keeps in touch with *Ellen*, whose family still farms the adjacent land. They still have coffee at City Limits Diner—the same booth, the same view of Michigan Avenue, the same waitresses who stopped asking what they wanted years ago. The diner sits across from Oakwood Cemetery, next to American Legion Post 322, a few hundred feet from Ford's old soybean plant. In the 1930s, Ford bought crops from seven hundred local farmers and processed them into oil for paint and plastics—part of his "Village Industries" program, where rural workers could draw factory wages without leaving their land. The plant closed after his death. The building became an antique shop, then an event venue. The dam still holds.

They sit in the same booth each time, the one by the window with the longest cathedral grains in the table. A framed aerial photo of old Saline hangs on the wall—the town before the subdivisions and the strip malls, when the farms ran unbroken to the horizon. The menu still runs from goulash to gluten-free. The coffee comes in heavy ceramic mugs that have outlasted most marriages. Semis carrying equipment pass on the same road that once carried Model Ts and covered wagons. They watch them go by, the way people in this booth have watched traffic go by for forty years.

Her well never ran dry. The consent agreement worked, at least for that. *Ellen* retired from teaching in 2027—thirty years at Saline High School, three decades of biology students who now work in hospitals and labs and, yes, data centers across the country. She can still identify every bird on her property by its song, though some songs she hears less often now. The cardinals stayed. The meadowlarks left when the fields did.

Her sons never developed any interest in farming. The older one works at the university, using artificial intelligence to find patterns in genomic data—she understood the genetics from thirty years of teaching, but not what the AI was doing with it. The younger one's company outgrew Ann Arbor; he moved to Detroit in 2027. His software reads contracts, flags issues, summarizes terms. He tried to explain it to her once, over Sunday dinner. She listened and nodded and thought about her former students who became paralegals. When the company raised its funding round, he was promoted. When AI tools started writing code as well as junior developers in 2029, his team shrank. He does not talk about work anymore when he visits. She does not ask.

Once, over Thanksgiving, after too much wine, he said something that stayed with her. "Mom, some of the people I work with—the serious ones, the ones who understand the math—they're not worried about losing their jobs. They're worried about losing everything. They have these probabilities they assign. P of doom, they call it. The chance that we build something we can't control." He laughed, but it wasn't a real laugh. "I don't know what number I'd put on it. But it's not zero. And I help build it anyway. Every day." She did not know what to say. She still doesn't.

The irony is not lost on her: her son builds tools that run on the same infrastructure she fought against, and those tools now threaten his own position—and perhaps, if the serious ones are right, threaten something larger still. Eventually the land will sell. She knows this now. She has made a kind of peace with it, the way you make peace with weather you cannot change. The philodendron her mother planted in 1978 still sits on her kitchen windowsill, in the same spot where her mother served breakfast for forty years. She waters it every Sunday morning, at the oak table scarred by a thousand family meals. Some things you can keep.

"She doesn't blame me," *Frank* says. "I asked her directly, a couple years ago. We were having coffee, watching the trucks go by, and I just asked. She said she understood. The township couldn't have won. The best we could do was get what we got." He pauses, turns his coffee cup in his hands, the honey-colored wood rail smooth against his elbow. The ceramic is warm. The coffee is cold. He drinks it anyway, sets it down—

Epilogue

the soft knock of ceramic meeting oak milled before he was born. "That's probably true. I hope it's true. I need it to be true."

Harold moved to Florida in 2027, to a condo near his daughter in Naples. The farm money set up his grandchildren for college, paid off debts he had carried for decades, bought him comfort he had never known. He died in 2029, quietly, in his sleep, with the air conditioning humming and palm trees outside his window instead of oaks. *Ellen* heard from his daughter. She called *Frank* that evening, and they sat together in the diner booth for an hour without saying much of anything, the booth backs' patchwork of dusty blues fading in the evening light. The waitress refilled their coffee twice without being asked. *Harold* had never come to those Sunday dinners she kept inviting him to. Now he never would.

"He understood," *Ellen* said finally. Her voice was rough. She had been crying before she got there; her eyes were red, and she kept pressing a napkin to them. "What we were losing. Even if he took the money. Especially because he took the money."

Frank nodded. He thought about that phone call years ago, when *Harold* had asked what he should do, and *Frank* had said he understood either way. He still meant it. He hoped *Harold* had known that, at the end. Outside the window, a semi-truck rumbled past, headed toward the facility. They watched it go.

David left the private equity fund in 2029. The returns had been strong—his investors were satisfied, his partners were wealthy—but something had shifted. He tells people he wanted to spend more time with his daughters, which is true. His oldest is in high school now, the one who showed him ChatGPT back in 2023 and changed his career. He missed too many of her volleyball games. He does not want to miss more.

The export control whiplash contributed too, though he does not talk about it much. Three policy reversals in 2025 alone. The H200 approval that December changed competitive dynamics overnight. By 2027, some facilities his fund had financed faced Chinese competition they had not modeled. The deals still worked, mostly. But the policy uncertainty wore on him. Building infrastructure for decades when policy shifts every administration felt like betting on a roulette wheel. He grew tired of recalculating.

He does not tell people about the sleepless nights, about the spreadsheets that stopped making sense. Somewhere along the way, the numbers had detached from anything real. A billion dollars became a rounding error. A thousand acres became a cell in a spreadsheet. The land had names once, families who worked it for generations. By the end, it was just coordinates and zoning classifications.

He thinks about *Frank* sometimes, though they never met. He read about the Saline Township consent agreement in the trade press. He recognized the pattern: local resistance, legal pressure, settlement, capitulation. His fund had been on the other side of similar negotiations. They had won, mostly. He is no longer sure what winning means.

"I still believe in what we built," he says, when asked. "The infrastructure matters. AI matters. But we could have done it differently. We could have listened more. We did not have to treat every community like an obstacle." He pauses. "I am not sure we knew how."

The buildout is not over. Decisions made now will shape what comes next; precedents set now will guide projects for decades. The questions that *Frank* and *Ellen* faced will confront other communities, in other states, with other resources and other vulnerabilities.

The miracle has faded from view. The costs have not.

Frank drives past the facility one more time. His wife asked him to pick up groceries, but he took the long way, the way that passes the property line. He takes Michigan Avenue, the old US-12—the road they called

Epilogue

the Chicago Road when pioneers first cut it through the wilderness in 1827, following Indian trails west toward a city that barely existed. His truck knows the route by heart. The same F-150 he has had for eleven years, 187,000 miles on the odometer.

The red barn catches afternoon light, the same light that has fallen on this land for centuries, long before his grandfather was born, long before anyone thought to build a metropolis of silicon on a cornfield. He remembers when it was just a barn surrounded by fields, *Harold*'s barn. The barn where *Harold* taught him to swap a filter, to cut down a PTO. Before anyone had heard of tokens or gigawatts or Stargate. Before the vote. Before the lawsuit. Before.

But now the buildings hum, not the tractors. *Frank* remembers the sound of *Harold*'s Deere from half a mile off. A diesel throb you felt in your chest before it reached your ears. This hum is different. Higher. Steadier. It never stops.

He thinks about what his father taught him. How you put seed in the ground and then you wait. You check the weather. You walk the rows. You watch for blight, for bugs, for the late frost that kills everything. And none of it is in your hands. You did the work. You spent the money. Now you wait to see if the world lets you keep what you planted.

That is what this feels like now. The concrete is poured. The copper is run. The chips are spinning. Somewhere, men in offices are watching numbers the way *Harold* used to watch the sky. Calculating yields. Praying for rain they cannot make. Hoping the harvest comes in before something fails. Before the money runs out. Before the weather turns.

Frank's father farmed eighty acres until 1974. He remembers helping bring in the corn, the exhaustion of it, the relief when the trucks pulled away full. He remembers the year the rain didn't come. His father walking the dying rows every evening, touching the curled leaves like he could will them back to life. There was nothing to do. The seed was in the ground. The money was spent. All you could do was wait and hope.

Nobody talks about the waiting. The months between planting and harvest when every storm makes your chest tight, every clear morning feels like borrowed time. Farmers know. You work and you spend and you hand it over to forces you cannot control. Weather. Markets. Luck. You find out in September whether you were right or wrong.

He watches the facility hum. Somewhere in those buildings, silicon is doing whatever silicon does. The seeds are different now. But the waiting is the same.

Somewhere, someone types a question.

But it does not matter if another question comes. It does not matter if the money men get their harvest in. The soil is already under concrete. The farms are gone.

This is server country now.

Selected Sources

This Essential Edition draws on the same primary sources as the full *This Is Server Country*: SEC filings, utility commission documents, press releases, industry reports, and investigative journalism. The full edition includes comprehensive source documentation with over 400 citations.

For readers seeking to explore the research in depth, the complete bibliography is available in the full edition and at `servercountry.org`.

www.ingramcontent.com/pod-product-compliance
Lightning Source LLC
Chambersburg PA
CBHW052129030426
42337CB00028B/5087